→INTRODUCING

EPIGENETICS

CATH ENNIS & OLIVER PUGH

This edition published in
the UK and the USA in 2017
by Icon Books Ltd,
39-41 North Road, London N7 9DP
info@iconbooks.com
www.introducingbooks.com

Sold in the UK, Europe and Asia
by Faber & Faber Ltd,
Bloomsbury House,
74–77 Great Russell Street,
London WC1B 3DA or their agents

Distributed in the UK, Europe and
Asia by Grantham Book Services,
Trent Road, Grantham, NG31 7XQ

Distributed in South Africa
by Jonathan Ball, Office B4,
The District, 41 Sir Lowry Road,
Woodstock 7925

Distributed in Australia and
New Zealand by Allen & Unwin Pty
Ltd, PO Box 8500, 83 Alexander
Street, Crows Nest, NSW 2065

Distributed in the USA
by Publishers Group West
1700 Fourth Street
Berkeley, CA 94710

Distributed in Canada
by Publishers Group Canada,
76 Stafford Street, Unit 300,
Toronto, Ontario M6J 2S1

Distributed in India
by Penguin Books India, 7th Floor,
Infinity Tower – C, DLF Cyber City,
Gurgaon 122002, Haryana

ISBN: 978-184831-862-5

Text and illustrations copyright © 2017 Icon Books Ltd

The author and artist have asserted their moral rights.

Editor: Kiera Jamison

Printed and bound in the UK by Clays Ltd, St Ives plc

Genes, RNA and Proteins

Epigenetics is about how the **genes*** we inherit from our parents are controlled, and how they interact with our environment: how our genes make us, well, *us*.

"Epi-" means upon, or in addition; epigenetics is the study of how additional factors interact with genes to direct the processes that make our cells* and bodies work.

Scientists have known about some of these factors for decades, but have only quite recently begun putting everything together to start explaining some of the gaps in our knowledge of genetics. From how embryos develop to how species evolve, from basic laboratory research to drug development – epigenetics is becoming a hot topic of conversation!

UNDERSTANDING EPIGENETICS – HOW OUR GENES INTERACT WITH OUR ENVIRONMENT AND OTHER FACTORS – IS CRUCIAL TO UNDERSTANDING MANY ASPECTS OF BIOLOGY.

* Words marked with an asterisk are defined in the glossary.

To understand epigenetics, we first need to know some basic genetics.

Genes are made of **deoxyribonucleic acid (DNA)***. DNA consists of long strings of four component molecules*, called **bases***: A, C, G and T. The order, or sequence, of these bases along the string serves as our genetic code.

Two long strings of DNA coil around each other to form the famous double helix structure. The bases on one strand form connections with the bases on the other strand; these connected pairs are the "rungs" in the twisted ladder-like structure of the helix. A always connects to T, and C always connects to G.

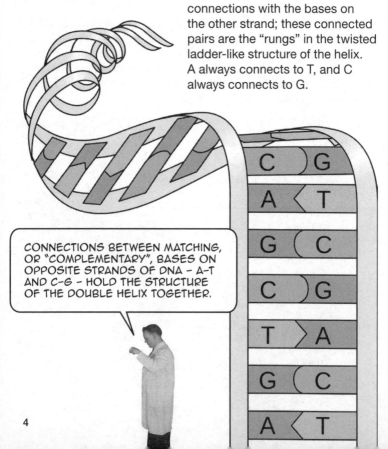

CONNECTIONS BETWEEN MATCHING, OR "COMPLEMENTARY", BASES ON OPPOSITE STRANDS OF DNA – A-T AND C-G – HOLD THE STRUCTURE OF THE DOUBLE HELIX TOGETHER.

C G
A T
G C
C G
T A
G C
A T

The first step in translating the DNA's coded instructions is called **transcription***. Part of the helix opens up, and the bases on one strand connect to new matching ("complementary") base molecules. The new bases link together into a strand of **ribonucleic acid (RNA)***. RNA is similar to DNA, but its short, single strands are less stable and more mobile than the DNA's long double helix.

Some types of RNA can squeeze out through tiny holes in the membrane that surrounds the cell nucleus*. DNA is too big to get through, so these RNA molecules act as coded messages from the genes to the rest of the cell.

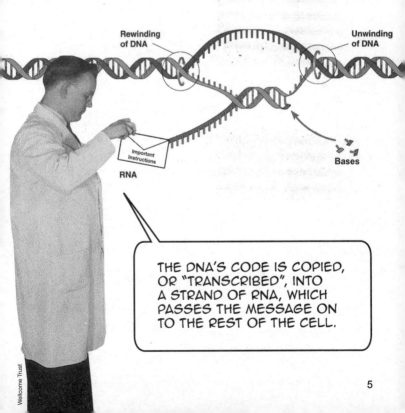

Rewinding of DNA

Unwinding of DNA

Important Instructions

Bases

RNA

THE DNA'S CODE IS COPIED, OR "TRANSCRIBED", INTO A STRAND OF RNA, WHICH PASSES THE MESSAGE ON TO THE REST OF THE CELL.

5

Some of the RNAs that leave the nucleus are called **messenger RNAs (mRNAs)***. mRNAs are copies of those sections of the DNA that code for large molecules called **proteins***.

Proteins are extremely important. There are thousands of different types, each with a specific function. Many proteins help to control the chemical reactions that keep our cells alive and healthy. For example, proteins are needed to open up the DNA double helix and to join individual bases together into RNA strands during transcription. Other proteins are involved in digesting food, fighting infections, carrying oxygen around the body, and thousands of other diverse functions.

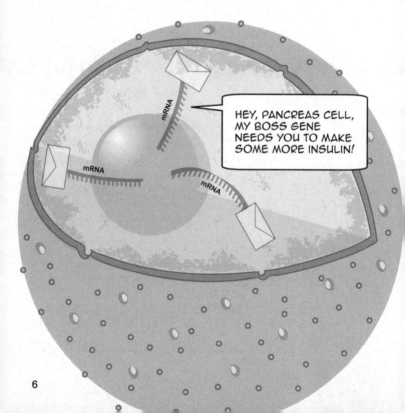

The process of converting mRNA sequences into proteins is called **translation***.

Each three-base unit – called a "**codon**"* – of mRNA connects to a **transfer RNA (tRNA)*** strand that has three complementary bases at one end. The other end is attached to a molecule called an **amino acid***. There are different types of amino acid, and each type can only attach to tRNAs that match specific codons.

Just as bases are the building blocks of DNA and RNA, amino acids are the building blocks of proteins. As tRNAs connect to their matching codons along an mRNA strand, their amino acids join up in the same order.

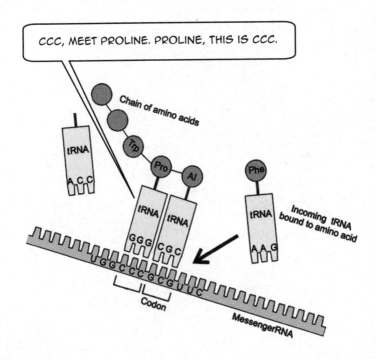

The sequence of amino acids in each protein is specified by the sequence of codons in the corresponding mRNA, which in turn matches the sequence of bases in the DNA. The very specific relationship between a given mRNA codon and its matching tRNA molecule, which is only ever attached to a single type of amino acid, is essential to the conversion of the DNA's code into proteins.

The sequence of amino acids in a protein determines its function. As we saw before, protein functions are essential for life. This is why DNA is so important – it contains all the instructions needed to make our cells and bodies work.

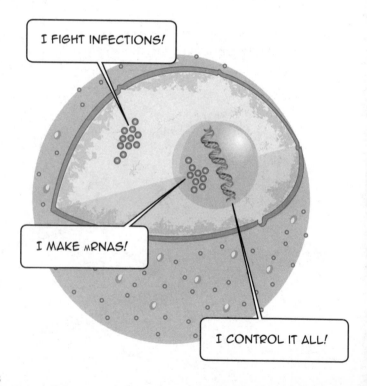

Chromosomes, Nucleosomes and Chromatin

Our complete DNA sequence is called our **genome***. All humans have extremely similar genomes, although we each have a slightly different version of the sequence. Almost every cell in your body contains its own copy of your unique version of the human genome.

The human genome is divided into 23 sections called **chromosomes***. Chromosomes come in pairs: we each inherit one chromosome of each pair from our mother, and the other from our father. The longest human chromosome contains about 2,600 protein-coding genes; the smallest, just 140. Genes are separated by stretches of non-protein-coding DNA.

THERE ARE TWO STRANDS OF DNA PER DOUBLE HELIX, AND ONE DOUBLE HELIX PER CHROMOSOME. THERE ARE 23 CHROMOSOMES FROM EACH PARENT, SO 46 CHROMOSOMES PER CELL. WHICH MEANS 92 STRANDS OF DNA PER CELL!

9

There are about 21,000 protein-coding genes in the human genome, which contains 3 billion individual bases (A, C, G and T). Laid out end-to-end, the DNA contained in a single cell would be about 1.8 metres (five feet) long. The DNA has to be twisted, folded and compacted to fit into a tiny cell nucleus.

The double helix first wraps around a cluster of eight small proteins called **histones***, which bind very tightly to DNA. Each individual unit of eight histone proteins plus DNA is called a **nucleosome***. The nucleosomes that assemble along a stretch of DNA look like beads on a string.

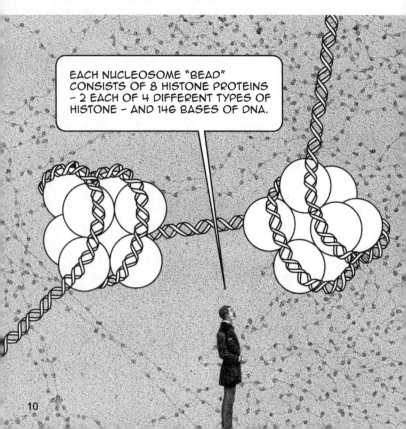

EACH NUCLEOSOME "BEAD" CONSISTS OF 8 HISTONE PROTEINS – 2 EACH OF 4 DIFFERENT TYPES OF HISTONE – AND 146 BASES OF DNA.

Four types of histone protein make up the nucleosome beads. A fifth type of histone protein attaches to the linker DNA between nucleosomes, and also to the histones inside each adjacent nucleosome. These connections compact the "beads on a string" into a thicker strand. Additional proteins, called scaffolds, bind to this strand and loop, fold and bend it into even denser structures.

The combination of DNA, histones and scaffold proteins, plus other proteins and RNAs that bind to the overall structure, is called **chromatin***.

The density of the chromatin varies along the length of each chromosome. You can actually see this variation in photographs of dyed cells – chromosomes are stripy, with the darker bands corresponding to regions of denser chromatin.

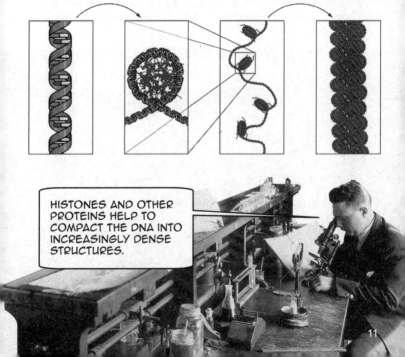

HISTONES AND OTHER PROTEINS HELP TO COMPACT THE DNA INTO INCREASINGLY DENSE STRUCTURES.

DNA Replication and Mitosis

New cells are constantly being created in our bodies, via a process called **cell division***.

Before a cell can divide in two, it has to make a second copy of its genome. This DNA replication process happens in a similar way to RNA transcription. The double helix opens up, breaking the connections between the two strands. The bases on each strand reconnect to complementary new partners, which link together into new strands of DNA.

The result at the end of the process is two double helices, each comprising one old strand of DNA from the original cell, and one newly formed strand.

Most cell divisions are of the type called **mitosis***, a process that creates two new cells which each have the same number of chromosomes as the original cell.

The chromosomes enter mitosis all jumbled up together like a bowl of spaghetti. As mitosis begins, they separate, condense and form pairs with their newly replicated copies. Fibres then extend out from opposite ends of the cell. Each chromosome attaches to a single fibre. As the fibres contract, one partner from each pair of chromosomes is pulled to each end of the cell. The cell membrane then pinches in at the middle to form two new cells, each surrounded by its own membrane.

Meiosis and Inheritance

Egg and sperm cells are created via a specialized form of cell division called **meiosis***. Meiosis involves two rounds of chromosome separation and cell division after DNA replication. Each of the four new cells created during a single meiosis event therefore receives 23 chromosomes, rather than the 23 *pairs* of chromosomes found in most other cells.

At conception, one egg and one sperm fuse to form a single cell. The 23 chromosomes inherited from each parent pair back up in this fertilized **zygote***, so that each new generation starts life with the same amount of DNA as its parents.

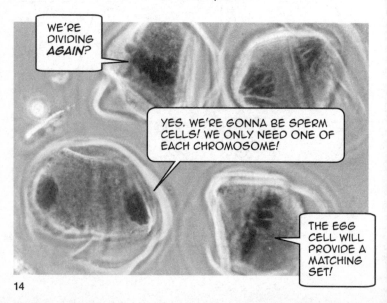

As chromosomes pair up for the first meiotic cell division, they swap segments of DNA with their partners. This genetic **recombination*** occurs when a chromosome breaks, and one of the two broken strands forms a double helix with the complementary sequence on the intact partner chromosome. The second intact strand is displaced by this intruder, and pairs instead with the other broken strand. Any gaps get filled in, and the pieces get stitched back together in their new locations. No information is lost – it's just remixed. The second round of meiotic cell division starts immediately after the first; no further recombination occurs.

HERE COMES THAT CRAZY CHROMOSOME REMIX! BUT I WON'T DROP THE BASE!

The number and locations of the chromosome breaks that trigger recombination are essentially random, so different pieces of DNA are swapped every time. This is why every sperm and egg cell is unique: they all get some of their DNA from each of the parent's two chromosomes, but in different combinations.

Unique eggs and sperm create unique offspring. You don't look exactly like your parents because their genetic material was shuffled before it was dealt to you; you don't look exactly like your siblings because their shuffle was different. The exception, identical twins, come from a single fertilized zygote that splits in two.

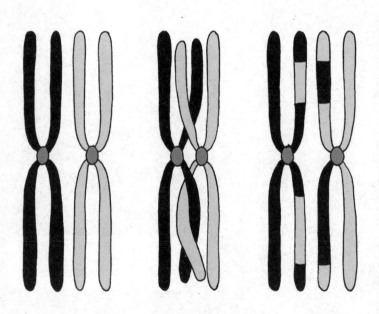

After conception, the fertilized zygote divides by mitosis to create all the different types of cells it will need as it matures into an embryo. This wonderfully intricate and complex process requires the zygote's genes to be transcribed and translated in carefully coordinated ways.

Twentieth-century biologists learned a lot about how this developmental process works, especially after the structure of DNA was characterized in 1953 by Watson, Crick, Wilkins and Franklin. However, as they put the pieces of the puzzle together, they also identified some gaps. Clearly, there are aspects of genetics that can't be explained by studying gene sequences alone.

Beyond the DNA Sequence: Gene Regulation

The human body comprises hundreds of different types of cell. Each type has specialized functions, mediated by its unique combination of proteins. Some proteins are produced in every cell; others are abundant in some cell types and present at low levels – or completely absent – in others.

The process by which the original fertilized zygote produces all of the body's cell types is called **cell differentiation***. Some cells differentiate as they undergo mitosis, becoming progressively more specialized, while others (called **stem cells***) stay in a less specialized, more versatile state. The cell's combination of proteins changes as it differentiates.

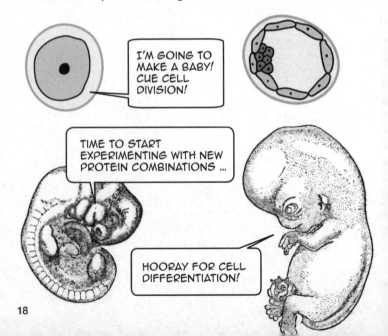

I'M GOING TO MAKE A BABY! CUE CELL DIVISION!

TIME TO START EXPERIMENTING WITH NEW PROTEIN COMBINATIONS ...

HOORAY FOR CELL DIFFERENTIATION!

Normal cell differentiation is a one-way process that converts versatile stem cells into more specialized mature cells. This ensures that mature brain cells, for example, don't spontaneously revert to being stem cells and start filling the skull with bone or muscle!

In 1962, **John Gurdon** (b. 1933) became the first scientist to artificially reverse cell differentiation. He took the nucleus of a fully differentiated tadpole gut cell and transferred it into a frog's egg from which the nucleus had been removed. The cloned egg matured into a new, healthy frog. This experiment proved that fully differentiated cells retain all the genetic material needed to produce every cell in the body.

A WHOLE, HEALTHY FROG, CLONED FROM A SINGLE MATURE GUT CELL! CELL DIFFERENTIATION *CAN* BE REVERSED! I COULD KISS YOU!

Gurdon's work disproved an earlier hypothesis that cells gradually discard unnecessary pieces of DNA as they differentiate, leaving behind only the genes they need to carry out their specialized functions. Modern science has since confirmed that with a few exceptions (some blood cells are weird), every cell in the human body contains exactly the same DNA as the original fertilized zygote. However, different cells transcribe and translate different parts of the genome, and it wasn't clear at the time how the same DNA sequence could be used to produce such diverse combinations of RNAs and proteins in different cell types.

HEY, BRAIN CELL, HOW ARE YOU MAKING PROTEIN C? I'VE GOT GENE C, BUT IT'S NOT ACTIVATED.

I DON'T KNOW. IT'S ONLY 1962!

In 1978, Hong Kong biochemist **Robert Tjian** (b. 1949) discovered the first example of a class of proteins called **transcription factors***, which help to regulate gene activation.

Transcription factors bind specific DNA sequences close to genes, and interact with the transcription machinery. The combination of proteins bound to any given gene helps to determine whether transcription takes place.

Some transcription factors are found only in certain cell types, and some are involved in cell differentiation. However, there isn't always a perfect correlation between the presence of a transcription factor and the activation of its target genes – sometimes transcription factors are present in cells where their target genes aren't activated. And it wasn't known how the production of the transcription factors themselves was regulated.

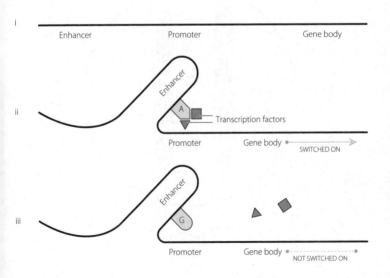

Some types of transcriptional regulation can't be explained by transcription factors alone, indicating that the cell must use additional mechanisms to control certain genes. One such example is called **imprinting***. Because chromosomes come in pairs, every cell contains two copies of each gene. Most genes are transcribed from both copies simultaneously. However, a few hundred imprinted genes are transcribed from only the maternally-inherited chromosome, and others from only the paternally-inherited chromosome.

Transcription factors can bind to either chromosome. Something other than transcription factors must therefore be involved in regulating imprinted genes.

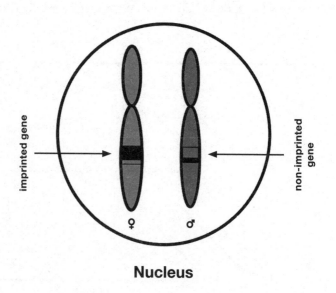

Nucleus

In some cells, a whole chromosome is completely shut down. Typically (although there are exceptions), female mammals receive two copies of the X chromosome – one from each parent – while male mammals receive one X chromosome from their mother, and a smaller Y chromosome from their father.

The X chromosome contains many more genes than the Y chromosome, and so XX cells have extra copies of some genes compared with XY cells. To compensate for this imbalance, one copy of the X chromosome becomes condensed and inactivated in every XX cell.

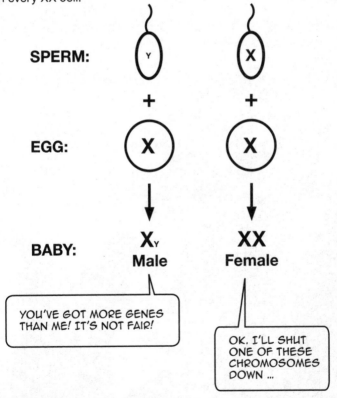

SPERM:

EGG:

BABY: X_Y
Male

XX
Female

YOU'VE GOT MORE GENES THAN ME! IT'S NOT FAIR!

OK. I'LL SHUT ONE OF THESE CHROMOSOMES DOWN ...

British geneticist **Mary Lyon** (1925–2014) discovered that, unlike imprinting, X chromosome inactivation is random: different cells shut down different copies.

The clearest demonstration of this randomness is seen in calico and tortoiseshell cats. Female cats (and the occasional XXY male) can inherit two different copies of a coat colour gene that's located on the X chromosome – one dark, one orange. Random inactivation of the chromosome that carries the orange gene causes dark colouration, and vice versa, resulting in a mosaic effect.

Like imprinting, X chromosome inactivation cannot be explained by transcription factors alone.

Nature and Nurture

The idea that our personal traits – collectively known as our **phenotype*** – are shaped by a combination of nature (our genes) and nurture (our environment, experiences and other non-genetic factors) isn't a new one. In fact, it predates **Gregor Mendel**'s (1822–84) discovery of the **laws of inheritance** by many centuries.

Mendel looked at how physical traits, such as height and flower colour, are inherited through several generations of pea plants. This work, combined with the later discovery of meiotic recombination, helped to explain nature's contribution to how we look and behave – why we're similar, but not identical, to our parents. However, the ways in which these genetic factors interact with our environment were much less clear.

TRAITS SUCH AS PLANT HEIGHT ARE INHERITED FROM BOTH PARENTS, BUT THEY DON'T BLEND – THEY ARE TRANSMITTED AS DISTINCT, UNCHANGING ENTITIES.

Twin Studies

There's more to us than just our genes: identical twins share the same DNA sequence, but have different personalities, preferences and medical histories.

Geneticists have learned a lot about nature and nurture by studying identical twins who were separated as babies and grew up in different families. For example, traits like facial features that are extremely similar in both twins are largely determined by genetics; traits that diverge, such as personalities, are more affected by the twins' different environments and experiences.

Because the separation of identical twins is a very rare scenario, most heritability* studies instead compare traits in pairs of identical and fraternal (non-identical) twins. Identical twins share the same environment *and* the same genome; fraternal twins have a common environment, but different DNA.

IT'S AMAZING THAT WE'RE BOTH TEACHERS. MAYBE THAT SKILL IS IN OUR GENES.

YEAH. THOUGH I NEVER REALLY GOT INTO ART, LIKE YOU. I GREW UP BEING MORE INTERESTED IN CHEMISTRY.

Twin studies have taught us a lot about complex traits that have both genetic and environmental components. For example, about 2.4% of all boys have an autism spectrum disorder. Boys whose *fraternal* twin brother has already been diagnosed have an increased risk – about 35%. The risk for boys whose *identical* twin brother is affected is higher again, at about 75%. The pattern is similar for girls, but the numbers are different. There's clearly both a strong genetic component (as the identical twins show us) *and* a role for environmental factors (as the fraternal twins demonstrate).

In 2015, an international team headed by **Peter Visscher** and **Danielle Posthuma** published the results of 50 years' worth of twin study data, concluding that, although the numbers vary between traits, the *average* breakdown is 49% nature, 51% nurture.

ALL THOSE TESTS, FOR OUR WHOLE LIVES ...

... AND THE ANSWER TO THE QUESTION OF HOW WE'RE AFFECTED BY NATURE AND NURTURE IS BASICALLY 50-50?!

THAT'S JUST THE AVERAGE! THERE'S A HUGE RANGE!

But what, exactly, is "nurture", and how does it work?

Much of our knowledge about DNA–environment interactions comes from research on complex diseases like cancer and heart disease, which have both genetic and environmental risk factors. This research has identified non-genetic factors that increase our risk of developing a disease (like smoking or poor diet), and others that protect us (like exercise). However, working out *how* these factors interact with our genes has proven more difficult.

The field of epigenetics can help to bridge this gap between the sequence of our DNA and the non-genetic factors that help to shape us.

EPIGENETICS CAN HELP TO EXPLAIN WHAT GENETICS ALONE CANNOT – INCLUDING HOW NATURE AND NURTURE WORK TOGETHER.

The History of Epigenetics

The original definition of epigenetics (or, rather, "epigenesis") concerned the mechanism of embryonic development. The Greek philosopher **Aristotle** (384–322 BC) proposed **epigenesis** – the gradual formation of an embryo from an amorphous starting point – as an alternative to **preformation**, which was the idea that an embryo grows from a preformed miniature version of itself.

By the mid-20th century, the crucial role that genes play in embryonic development was widely acknowledged. The updated epigenesis model of the time proposed that interactions between the genetic material, proteins and as yet unknown chemicals drive embryonic development, producing all the different cells and tissues of the body.

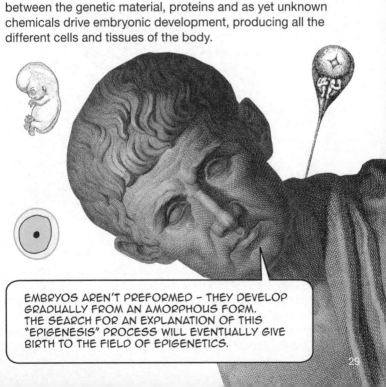

EMBRYOS AREN'T PREFORMED – THEY DEVELOP GRADUALLY FROM AN AMORPHOUS FORM. THE SEARCH FOR AN EXPLANATION OF THIS "EPIGENESIS" PROCESS WILL EVENTUALLY GIVE BIRTH TO THE FIELD OF EPIGENETICS.

One of the proponents of the updated epigenesis theory of embryonic development was the British developmental biologist **Conrad Waddington** (1905–75). In a paper published in 1942, Waddington merged the older term "epigenesis" with "genetics" to create the new word "epigenetics".

'EPIGENETICS IS THE BRANCH OF BIOLOGY WHICH STUDIES THE CAUSAL INTERACTIONS BETWEEN GENES AND THEIR PRODUCTS [PROTEINS] WHICH BRING THE PHENOTYPE INTO BEING.'

Waddington recognized that different cells must contain different epigenetic features, which he called "landscapes", and that cell differentiation involves a change in this landscape. He likened the one-way cell differentiation process to a ball rolling downhill through the epigenetic landscape as the embryo develops. Once it has rolled downhill, or differentiated (into, for example, skin or liver cells), it is incredibly difficult to roll it back up to its undifferentiated (stem cell) state.

In 1958, American geneticist **David Nanney** (b. 1925) used the term "epigenetics" in a slightly different way, referring to aspects of biology that cannot be explained by genetics alone. Nanney's paper sparked decades of debate about the true definition of the word, and skirmishes still erupt today.

'CELLS WITH THE SAME GENOTYPE [GENES] MAY NOT ONLY MANIFEST DIFFERENT PHENOTYPES, BUT THESE DIFFERENCES IN EXPRESSED POTENTIALITIES MAY PERSIST INDEFINITELY DURING CELLULAR DIVISION IN ESSENTIALLY THE SAME ENVIRONMENT.'

The idea of "persistence during cell division" refers to the fact that mature cells divide by mitosis to form two cells of the same type (liver cells only make more liver cells). The two new cells have the same epigenetic landscape and the same phenotype as the original cell. This persistence could explain the one-way nature of cell differentiation: if epigenetic landscapes don't change when a mature cell divides, the cell cannot dedifferentiate.

Most disputes about the definition of epigenetics concern whether the persistence of epigenetic landscapes during mitosis is an essential part of the description. This persistence is also commonly called "mitotic heritability", because both cells produced during mitosis inherit genes and epigenetic landscapes from the original cell. (The concept is similar to, but distinct from, the more familiar type of inheritance from parent to child.)

More conservative modern definitions state that "epigenetics" should refer only to mitotically heritable factors that interact with our genes. This book uses a looser interpretation, so as to encompass additional factors that don't necessarily persist during cell division. People will argue about this.

MITOTIC HERITABILITY IS CRUCIAL TO MAINTAINING EPIGENETIC LANDSCAPES.

BUT THERE'S MORE TO MODERN EPIGENETICS THAN THAT!

The Discovery of Chromatin Modifications

We've already discussed how chromatin comprises both DNA and proteins. Up until the mid-20th century, it was widely believed that the proteins must be the important part. Surely DNA, with its simple structure and only four different bases, couldn't possibly be responsible for building a complex organism from scratch.

However, in the 1950s, American geneticists **Martha Chase** (1927–2003) and **Alfred Hershey** (1908–97) used purified viral DNA to prove that the inherited instructions for life are indeed coded into the DNA sequence. Histone and scaffold proteins fell out of favour, regarded simply as packaging for the genetic material. It wasn't until several decades later that histones started to become interesting again.

ACTUALLY, IT LOOKS LIKE DNA IS THE BASIS OF ALL LIFE.

Scientific interest in histones was revived in the mid-1990s when American developmental biologist **David Allis** (b. 1951) started to discover chemical changes to histone proteins – changes that correlate with gene transcription.

Specifically, Allis's team found that a small molecule called an **acetyl group*** is added to histones in some regions of the genome, and that these regions contain more actively transcribed genes than regions with very few acetylated histones. Allis and others went on to find several other types of molecular histone modifications, some associated with active parts of the genome, and others associated with transcriptionally silent regions.

HISTONE ACETYLATION
CORRELATES WITH
THE TRANSCRIPTION
OF GENES. HISTONES
ARE BECOMING
INTERESTING AGAIN!

The physical configuration of the chromatin structure also correlates with gene transcription: more open, lighter chromosome "stripes" generally contain more active genes than do densely packed, dark "stripes".

Some parts of the genome are always found in a dense, inactive chromatin state. The ends and central junctions of chromosomes fall into this inactive category. Other regions, for example those that contain lots of genes involved in essential processes that are needed in every cell (such as copying DNA or converting sugars into energy), are always found in a more open configuration. Other regions have different chromatin densities in different types of cell – suggesting an association with cell-specific gene regulation.

> THE ENDS AND CENTRAL JUNCTIONS OF CHROMOSOMES, AND THE OTHER DARK STRIPES, CONTAIN DENSE CHROMATIN AND ARE LARGELY SILENT. THE LIGHTER STRIPES CONTAIN MORE OPEN CHROMATIN AND ACTIVE GENES.

Some researchers were studying chemical changes to the DNA itself. In 1948, American biochemist **Rollin Hotchkiss** (1911–2004) demonstrated that some DNA bases have small molecules called **methyl groups*** attached to them. However, the function – if any – of **DNA methylation** wasn't known.

Inactivated X chromosomes were known to contain many methylated bases. In the mid-1970s, British geneticist **Adrian Bird** (b. 1947) started to find evidence that the methylation of other DNA regions might also be associated with transcriptional regulation; sites that can be methylated aren't scattered throughout the genome at random, but correlate with the locations of genes. Americans **Mark Tykocinski** and **Edward Max** and Belgian **Benoit de Crombrugghe** published similar findings in the early 1980s.

DNA METHYLATION SEEMS TO BE INVOLVED IN TRANSCRIPTIONAL REGULATION – IT CORRELATES WITH TRANSCRIPTIONAL SILENCING.

Chromatin modifications at last seemed like promising candidates to bridge the gap between the sequence of the DNA and the phenotype (observable characteristics) of the whole organism. However, it was first necessary to establish cause and effect.

DO CHROMATIN MODIFICATIONS CONTROL GENE TRANSCRIPTION?

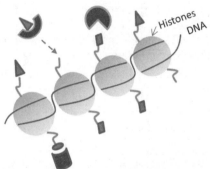

OR ARE CHROMATIN MODIFICATIONS ADDED TO REGIONS OF THE GENOME WHERE TRANSCRIPTION HAS ALREADY BEEN ACTIVATED OR SILENCED BY OTHER MECHANISMS?

It's easy to see how chromatin density can control gene transcription directly: denser structures physically block the transcription machinery from reaching the DNA. However, the potential roles of DNA and histone modifications in transcriptional regulation were less obvious.

In the early 1980s, both **Carol Prives** in the USA and **Walter Doerfler** (b. 1933), leading a team of German, Swiss and Israeli scientists, artificially methylated pieces of viral DNA and injected them into cells. Both teams found that unmethylated DNA is more transcriptionally active than methylated DNA with the same sequence.

In the 1990s, David Allis demonstrated that proteins which add acetyl groups to histones are directly responsible for activating gene transcription in some situations.

These findings laid the groundwork for an exciting few years of research. As evidence accumulated, it became obvious that DNA and histone modifications do indeed directly help to control gene transcription.

The Modern Understanding of Epigenetic Modifications

We now know that DNA and histone modifications represent additional layers of information superimposed onto the DNA sequence. Modern-day epigenetics is the study of this "information" and:

- How it is established, maintained and modified;
- How its code is translated by the cell;
- How it is inherited, both in the short term by the next generations of cells and organisms, and over the much longer periods of time over which evolution takes place;
- How it becomes distorted and scrambled in disease; and
- How we can read and perhaps learn to edit this information to help improve our health.

If the DNA sequence is an instruction manual that explains how to make a whole organism from a fertilized zygote, then epigenetic information is a highlighted and annotated version of the text.

Some molecular "colours" denote the parts of the text that need to be read most carefully, and others mark parts that can be ignored. The highlighting helps to determine which genes are transcribed and translated in which cells.

The cell uses various RNAs and proteins as epigenetic "highlighters" that establish and maintain the pattern of information, "erasers" that remove it when necessary, and "decoders" that convert the information into usable instructions.

THIS CERTAINLY MAKES THE DNA A LITTLE EASIER TO WORK WITH.

There are thousands of epigenetic highlighters, erasers and decoders, working together in a complex and carefully coordinated network.

The epigenetic regulation network encompasses diverse types and sizes of molecules. The smallest known epigenetic modification consists of only four atoms; RNA strands that start at just nineteen bases long help to specify where each type of highlighting should go. Proteins of various sizes add, remove or recognize specific modifications, while others move individual nucleosomes around. Whole long sections of chromosomes are corralled into specific areas of the cell's nucleus. There are, no doubt, additional components that we haven't even identified yet.

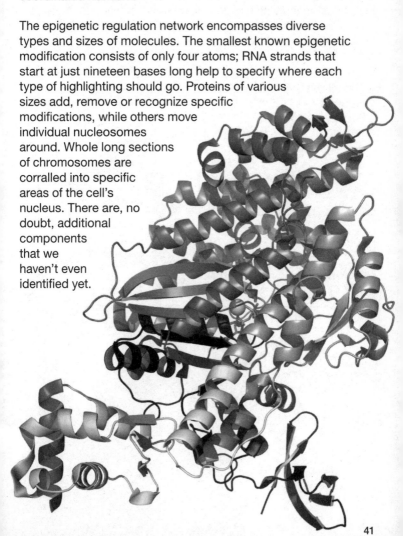

Almost every cell in the body contains the same DNA sequence, but different cell types contain different patterns of molecular highlighting. A liver cell doesn't need the same pages of the instruction manual as a brain cell, after all. Instead, each cell produces only the RNAs and proteins that it needs to carry out its specialized functions.

One of the really interesting things about epigenetics is that the marks aren't fixed in the same way the DNA sequence is: they change as cells differentiate, and in response to certain outside influences. Some might even be passed on from parent to offspring.

I NEED DIFFERENT HIGHLIGHTING PATTERNS SO I CAN MAKE SOME NEW PROTEINS TO DEAL WITH THIS ALCOHOL.

DNA Methylation

The smallest known epigenetic modification is called a methyl group. This molecule, comprising one carbon atom and three hydrogen atoms, is attached to some of the DNA's C bases by a DNA methyltransferase "highlighter" protein.

DNA methylation causes gene silencing. Adrian Bird discovered decoder proteins in the cell nucleus that recognize and bind specifically to methylated C bases (mC). These proteins shut down transcription from genes that contain methylated DNA, preventing the production of the corresponding RNAs and proteins.

DNA methylation is really more like a censor's black ▇▇▇▇▇ than a highlighter, telling the cell: "nothing to see here".

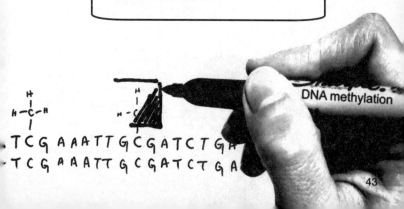

GENES THAT CONTAIN METHYLATED C BASES ARE SHUT DOWN BY DECODER PROTEINS THAT BIND TO THE METHYL GROUPS AND BLOCK TRANSCRIPTION.

DNA methylation

T C G A A A T T G C G A T C T G A
T C G A A A T T G C G A T C T G A

When Adrian Bird and others first started to map DNA methylation patterns in the 1970s, the methods they used were crude and slow. In the 1990s, though, Australian geneticists **Marianne Frommer** and **Susan Clark** developed a method that can reveal exactly which C bases are methylated and which are not.

Before bisulphite conversion →

T C G A A G C̲ G T A C̲ A

After bisulphite conversion →

T C G A A G T G T A T A

A CHEMICAL CALLED *SODIUM BISULPHITE* CONVERTS UNMETHYLATED C BASES INTO TS, BUT METHYLATED C BASES ARE PROTECTED AND REMAIN AS CS. THE METHYLATION PATTERN IS READ BY COMPARING THE SEQUENCE OF THE BISULPHITE-TREATED DNA TO THAT OF THE STARTING DNA.

Bisulphite sequencing experiments are now used by researchers all over the world to help map the epigenetic landscapes of different types of cells.

DNA methylation doesn't occur at random, but rather follows some general rules and patterns.

Most methylated Cs are adjacent to G bases. Many active genes have a cluster of these around their transcription start sites. These cluster features are called **CpG islands***, and most are unmethylated. In contrast, CpGs between genes, or in **repetitive DNA sequences***, are usually methylated.

Repetitive DNA is problematic. It can move to new locations; it can make the transcription and DNA replication machinery slip and stumble, causing mutations that contribute to cancer and other diseases. It's possible that DNA methylation originally evolved as a way to shut these troublesome sequences down, and acquired its other useful functions later.

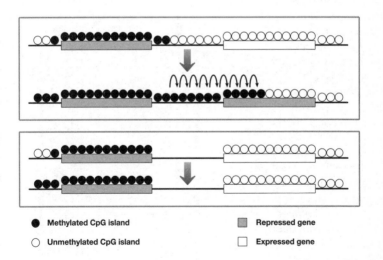

● Methylated CpG island   ▨ Repressed gene

○ Unmethylated CpG island   ▢ Expressed gene

Transcriptional silencing only occurs when both strands of the DNA double helix are methylated at the same CpG site. However, when DNA is replicated during mitosis, the newly formed strands that pair with each of the two original strands have no methyl groups attached.

Most of the time, both of the cells produced during mitosis need to retain the same DNA methylation patterns that were present in the original cell. In other words, the epigenetic landscape needs to persist, to ensure that both new cells are of the same type as the original (see pages 31–2). The DNA methylation pattern from each original DNA strand therefore has to be copied onto each newly formed strand.

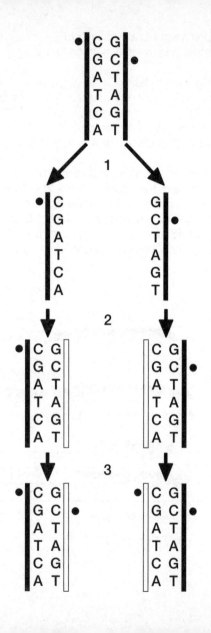

46

A protein called DNA methyltransferase 1, or DNMT1, is responsible for copying the original DNA methylation pattern to the newly formed strands. DNMT1 recognizes and binds specifically to CpG sites that are asymmetrically methylated. It then adds a methyl group to the naked C base on the new DNA strand, restoring the original cell's symmetrical methylation pattern. This process is crucial to maintaining the epigenetic landscapes of mature cells and preventing the reversal of cell differentiation.

DNA methylation is the best-known example of a mitotically heritable epigenetic modification (see pages 31–2). Even the strictest definitions of epigenetics include DNA methylation!

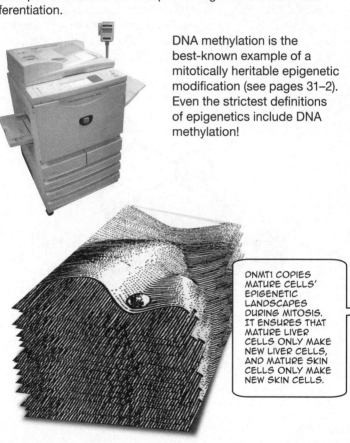

DNMT1 COPIES MATURE CELLS' EPIGENETIC LANDSCAPES DURING MITOSIS. IT ENSURES THAT MATURE LIVER CELLS ONLY MAKE NEW LIVER CELLS, AND MATURE SKIN CELLS ONLY MAKE NEW SKIN CELLS.

Sometimes though, methylated genes have to be reactivated – for example during cell differentiation. Epigenetic landscapes change during this process, as cells start to activate the genes they will need to perform the specialized functions of a mature brain, liver, blood or kidney cell.

In most of these situations, the original methylation pattern is simply not copied to the newly formed DNA strands. This gradual dilution of the methylation pattern is called passive demethylation. Because **passive demethylation** is dependent on DNA replication, it can only be used to reactivate silenced genes in cells that are dividing by mitosis.

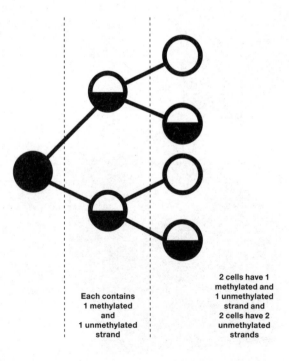

Each contains 1 methylated and 1 unmethylated strand

2 cells have 1 methylated and 1 unmethylated strand and 2 cells have 2 unmethylated strands

Methylation sometimes needs to be reversed very quickly – for example, in early-stage embryos undergoing rapid cell differentiation. Sudden gene reactivation is also sometimes needed in mature cells that aren't currently dividing – for instance, in response to chemicals, temperature changes or other stimuli. Passive dilution of methylation during cell division is unsuitable for these occasions; a separate, active process is needed.

During **active demethylation**, the methyl groups that need to be removed are tagged with oxygen atoms. "Eraser" proteins called Tet bind specifically to the tagged methyl groups and snip them off the DNA.

> ERASERS ARE JUST AS IMPORTANT AS HIGHLIGHTERS: THEY ALLOW CELLS TO CHANGE THEIR GENE ACTIVATION PATTERNS DURING CELL DIFFERENTIATION AND IN RESPONSE TO CHANGES IN CONDITIONS.

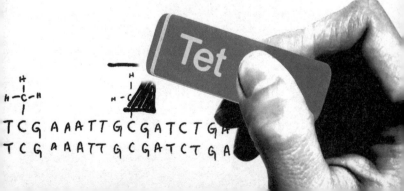

In the 1990s, an American team led by German biologist **Rudolf Jaenisch** (b. 1942) demonstrated the importance of DNA methylation by genetically engineering mice that lack DNA methyltransferase proteins. The mice died as embryos.

Several other teams of researchers have found that the general chaos within human cancer cells includes massive changes in DNA methylation and gene activation patterns (see page 136). Mutated DNA methyltransferase proteins have also been found in some types of cancer cells.

These and other studies suggest that normal DNA methylation patterns are needed for normal cell and organism function. However, DNA methylation doesn't work alone; other forms of epigenetic modification also help to control gene activation.

'DNA METHYLATION IS ABSOLUTELY ESSENTIAL FOR THE DEVELOPMENT AND HEALTH OF THE ORGANISM.'

Histone Modifications

Like DNA, histone proteins can be tagged with methyl groups, and also with a plethora of other molecules, each with different functions. Most modifications are added to the histone "tails" – the parts of the proteins that poke out of the core of the nucleosome structure. Histone modification patterns change more often, and more quickly, than DNA methylation patterns. In general, they seem to be associated with short-term fluctuations in gene activation patterns, rather than the longer-term changes mediated by DNA methylation.

Just as there are proteins that bind specifically to methylated DNA to shut down gene transcription, there are decoder proteins that bind specifically to each type of histone modification and regulate the activity of nearby genes.

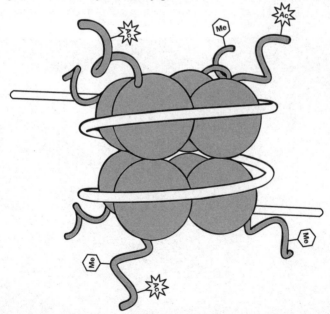

Because DNA methylation affects specific individual C bases, it's relatively easy to identify the precise location of every methyl group. Each nucleosome, on the other hand, is associated with about 150 bases worth of DNA, plus the 80-base linker sequence.

An indirect method called **ChIP-Seq*** (for "Chromatin Immunoprecipitation Sequencing") is used to determine which parts of the genome are associated with which kinds of histone modifications – the first step to understanding what the modifications do.

ChIP-Seq involves isolating histones that are tagged with a specific kind of modification, and sequencing the associated DNA to identify its location in the genome.

Y-SHAPED PROTEINS CALLED ANTIBODIES CAN BIND SPECIFICALLY TO A SINGLE TYPE OF HISTONE MODIFICATION, AND SEPARATE THE MODIFIED CHROMATIN FROM THE REST OF THE GENOME FOR ANALYSIS.

DNA methylation is always associated with gene silencing. In contrast, histone methylation can either silence or activate genes, depending on which amino acids of which histone tails are methylated, and whether an amino acid receives one, two or three methyl groups. Every configuration attracts and binds different decoder proteins.

Different types of histone methylation identify parts of the genome with different characteristics. Some types of methylation define active parts of the genome, and others, the silent regions. Other configurations mark damaged DNA that's being repaired, or pieces of DNA that help to control nearby or distant genes.

REGIONS OF THE GENOME THAT ARE ACTIVE, SILENCED OR UNDERGOING DNA REPAIR ARE MARKED WITH DISTINCTIVE HISTONE METHYLATION PATTERNS.

Other molecules can also attach to histone tails. Acetyl and phosphate groups are two of the smallest – about the same size as methyl groups. **Histone acetylation** was the first histone modification to be discovered, by David Allis (see page 34). It is always associated with gene activation. Not only do transcription-activating decoder proteins bind to acetylated amino acids, but acetyl groups also have a more direct effect: their negative charge weakens the attraction between the negatively-charged DNA and the positively-charged histones. This interference relaxes the structure of the nucleosome, making the DNA easier to transcribe. **Histone phosphorylation**, another modification, is less well understood, but is associated with DNA repair and transcriptional activation.

Additional histone modifications have been discovered more recently, including ADP-ribose molecules, which are one of the larger histone modifications. **Histone ADP-ribosylation** seems to act in a similar way to acetylation, physically disrupting the nucleosome structure to make the DNA easier to transcribe. Certain proteins, which are larger still, can also be attached directly to histone tails. The attachment of SUMO and ubiquitin proteins seems to be associated with both gene silencing and gene activation, depending on the attachment site.

Identifying, and then understanding, additional histone modifications is still a very active and ongoing area of research. The list of known histone modifications, and their known and possible roles, seems to grow longer every year!

The most drastic histone modification involves swapping a standard histone for a variant protein with specialized properties. Some **histone variants*** make the nucleosome more stable, impeding transcription; others have the opposite effect. Some contain modifiable amino acids that don't exist in standard histone types; some are thought to be involved in repairing damaged DNA.

HISTONE VARIANTS CAN SUBSTITUTE FOR THE STANDARD HISTONE PROTEINS WHEN NEEDED. A VARIANT CALLED H2A.X REPLACES HISTONE H2A IN DAMAGED PARTS OF THE GENOME THAT NEED REPAIR.

In sperm cells, histones are completely removed from most of the genome and replaced with proteins called protamines, which are smaller than histones and can pack DNA into an extremely compact, inactive form – necessary in such a small cell. This substitution also allows the fertilized zygote to keep track of which of its chromosomes came from the sperm and which from the egg, which is important for proper embryonic development.

Unlike DNA methylation, it was long thought that histone modification patterns weren't copied directly to the new chromosomes produced during mitosis. As a result, histone modifications have traditionally been excluded from more conservative definitions of epigenetics, which require the persistence of epigenetic features during mitosis.

However, a 2014 study by American developmental biologist **Susan Strome** showed that some of the original strand's modified histones are passed to the newly forming strand during DNA replication. The modification patterns then spread to the fresh, unadorned nucleosomes that form on both strands.

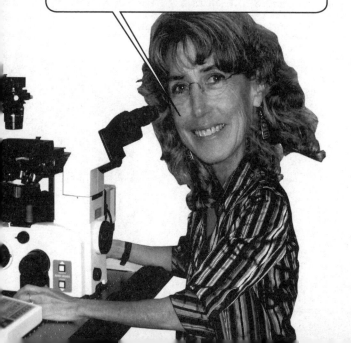

'THERE HAS BEEN ONGOING DEBATE ABOUT WHETHER THE METHYLATION MARK CAN BE PASSED ON THROUGH CELL DIVISIONS, AND WE'VE NOW SHOWN THAT IT IS.'

Many researchers are still working out the function of each histone modification. But even when they do eventually characterize each individual modification, they still won't have the complete picture; they also need to understand how every possible *combination* of histone modifications works.

Some context-dependent effects have already been characterized. For example, some types of histone methylation only help to activate nearby genes in regions of chromatin that also contain acetylated histones. This type of interaction allows for incredibly precise, tightly-regulated control over gene transcription. However, we've still barely scratched the surface of the complete histone code.

THE HISTONE CODE IS MUCH MORE COMPLICATED THAN DNA METHYLATION. IT CONTAINS MULTIPLE TYPES OF MODIFICATIONS, AND COMBINATIONS THEREOF, EACH WITH DISTINCT FUNCTIONS.

Chromatin Remodelling

Nucleosomes aren't fixed in place – they can slide along the DNA. The carefully coordinated disassembly, reassembly and movements of nucleosomes are important components of epigenetic regulation.

The proteins that coordinate this chromatin remodelling process were discovered in yeast cells, where they were first thought to have very specialized roles. For example, SWI/SNF remodellers (versions of which are also present in human cells) were named for "Mating Type Switch/Sucrose Non-Fermenting" after their very yeast-specific roles! Since then, however, roles for chromatin remodelling proteins have been identified in many essential processes, such as DNA replication and embryonic development, in humans and other complex organisms.

CHROMATIN REMODELLING PROTEINS HELP TO REPLICATE DNA, AND TO CONTROL GENE TRANSCRIPTION AND OTHER PROCESSES, IN YEAST AND HUMAN CELLS.

Chromatin remodelling proteins help to regulate the spacing between nucleosomes. Moving nucleosomes closer together creates stronger contacts between the histone proteins, which condenses the chromatin into a very compact form; spacing them out has the opposite effect, opening up the chromatin into a more accessible, active configuration.

An accessible configuration can be highly useful, for example, to allow the proteins that replicate, transcribe or repair the DNA to access the double helix without interference from histones. In this case, the nucleosomes may even be completely removed. Breaking the attraction between the positively-charged histones and negatively-charged DNA requires a lot of energy.

Chromatin remodelling proteins help to incorporate histone variants with specialized properties, such as DNA repair, into nucleosomes (see page 56). This substitution process requires the nucleosomes to be disassembled and then reassembled.

Nucleosomes that contain only the standard, non-variant histone proteins are also sometimes stripped down and rebuilt using a mix of fresh and modified histones. This process might be used as a kind of shortcut when histone modification patterns need to change more rapidly than usual – for example, during cell differentiation or in response to sudden changes in the environment.

LET'S REMOVE ALL THESE HISTONE MODIFICATIONS AND ADD DIFFERENT ONES.

HOLD ON. REPLACING THE WHOLE HISTONE PROTEIN WOULD BE FASTER.

Nuclear Location

A cell's nucleus has distinct "neighbourhoods" that specialize in different functions. For example, active regions of chromatin, which contain epigenetic marks that promote gene transcription, congregate together in the equivalent of a downtown entertainment district; silenced regions of chromatin stay away, confined to quieter residential suburbs.

Each type of mature cell has a distinctive pattern of gene locations within the nucleus, reflecting its unique epigenetic landscape and gene activation pattern. This means that many of the genes in a liver cell are found in different parts of the nucleus compared with the same genes in a brain cell.

ACTIVE GENES CONGREGATE IN TRANSCRIPTION "HOT SPOTS" IN THE CENTRAL REGION OF THE CELL NUCLEUS. EACH SPOT SPECIALIZES IN TRANSCRIBING GENES THAT ARE ACTIVATED AT THE SAME TIME OR UNDER THE SAME CONDITIONS.

RNA

Not all of the RNA strands copied from the DNA template during transcription are translated into proteins; some have distinct functions of their own, including roles in the epigenetic regulation of gene activation.

Single strands of RNA of any length can pair up with DNA or other RNA strands that contain even very short complementary sequences. Longer RNA strands can also bind to complementary parts of themselves, allowing them to fold up like origami into 3D shapes called "secondary structures", which can be recognized by certain proteins. RNA molecules use complementary pairing and secondary structures to help coordinate which epigenetic modifications should be made to which parts of the genome.

FOLD IT INTO A LOOP, MAKING SURE THAT THE BASE PAIRS ARE COMPLEMENTARY ALONG THE STEM. OK, NICE SECONDARY STRUCTURE!

The longest and most complex RNA molecules that can coordinate epigenetic regulation have the imaginative name of **long non-coding RNAs**, or **lncRNAs***.

lncRNAs are at least 200 bases long and have various functions within the nucleus. One important role is to help direct where each type of epigenetic modification should be attached to the chromatin.

The way lncRNAs do this is by inserting one end of the strand into the double helix, binding to a complementary DNA sequence. Various highlighter and eraser proteins can bind to the secondary structures formed by the other end of the lncRNA strand; these proteins then modify and remodel nearby chromatin.

lncRNA

SORRY FOR INTERRUPTING, BUT I HAVE A DELIVERY OF EPIGENETIC MODIFICATIONS FOR THIS ADDRESS. I NEED A SIGNATURE PLEASE.

Some RNAs shut down transcription from destabilizing repetitive DNA sequences (see page 45). **PIWI-interacting RNAs (piRNAs)** leave the nucleus after being transcribed and bind to a type of protein called PIWI, which they then bring back into the nucleus. The piRNAs are 26 to 31 bases long and bind to complementary sequences in RNAs that are being actively transcribed from repetitive DNA. This marks the new RNA strands for destruction. The attached PIWI proteins simultaneously recruit partners that methylate nearby DNA, which prevents any further transcription from taking place.

lncRNAs and piRNAs can stay associated with their respective DNA sequences during cell division.

IF WE DON'T STOP REPETITIVE DNA FROM MOVING AROUND THE GENOME IT WILL CAUSE HARMFUL GENE MUTATIONS.

Like lncRNAs, the smallest known regulatory RNAs have a very imaginative name: **microRNAs**, or **miRNAs***. Longer precursor strands of RNA are trimmed down to 19 to 24 bases after leaving the nucleus. Mature miRNAs can then exert their effects by binding to complementary mRNA molecules, preventing their translation into proteins. Perfectly matched miRNA sequences trigger mRNA strand destruction, while imperfectly matched miRNAs block the translation machinery. An individual miRNA can target many different mRNAs, and translation of an individual mRNA can be blocked by many different miRNAs.

American geneticists **Craig Mello** (b. 1960) and **Andrew Fire** (b. 1959) discovered mRNA silencing by short complementary RNAs in 1998. The discovery had an immediate impact: miRNAs are now widely used as research tools. Targeting an individual mRNA with miRNAs can reveal the role the corresponding protein plays in the cell. For example, if adding a miRNA makes the cell stop dividing, then the corresponding protein might be involved in mitosis. It's much easier to inject a miRNA into a cell to prevent a protein from being translated than it is to delete the corresponding gene.

Mello and Fire's work won the 2006 Nobel Prize in Physiology or Medicine – one of the shortest ever gaps between discovery and prize.

'THIS YEAR'S NOBEL LAUREATES HAVE DISCOVERED A FUNDAMENTAL MECHANISM FOR CONTROLLING THE FLOW OF GENETIC INFORMATION.'

There are many thousands of lncRNAs, piRNAs and miRNAs, with more still being discovered each year. In 2013, German and Danish teams led by **Nikolaus Rajewsky** (b. 1968) and **Jørgen Kjems**, respectively, independently discovered that circular RNAs – previously thought to have no function – can also regulate transcription, by mopping up miRNAs so that they can't block protein translation.

Most regulatory RNAs are produced only in certain types of cell, at certain stages of development or in response to changes in the cell's environment, such as a bacterial infection. The specific combination of regulatory RNAs present in each cell helps to define which genes are transcribed and which proteins are made.

'THERE SEEMS TO BE A WHOLE NEW LAYER OF GENE REGULATION.'

'THE MOLECULES COMPRISE "A HIDDEN, PARALLEL UNIVERSE" OF UNEXPLORED RNAS.'

Interactions Between Different Epigenetic Modifications

We've seen how lncRNAs and piRNAs exert their effects on gene transcription by recruiting proteins that can modify DNA, histones and larger chromatin structures. We've also seen how the effects of some histone modifications depend on which other modifications are present in the same region of chromatin (see page 58).

Other types of epigenetic modifications interact with each other, and with transcription factors, in similar ways. These interactions can reinforce or fine-tune the regulation of gene transcription, and can allow active or silent chromatin states to spread from their initial location to adjacent genes and beyond.

ALMOST THERE!

WE'VE GOT TO WORK TOGETHER TO SHUT THIS GENE DOWN!

The decoder proteins that bind specifically to methylated DNA and other repressive epigenetic marks have another trick up their sleeves: they can recruit additional highlighter proteins. The new recruits methylate nearby CpGs, attach repressive histone modifications or remodel the chromatin into a denser state. The decoders that recognize acetyl groups and other transcription-activating histone modifications amplify and reinforce each other's signals in a similar way.

The order in which epigenetic modifications reinforce each other isn't clear yet – we don't know which modification establishes the repressive or activated state, and which maintains it. It probably depends on the situation.

The RNAs and proteins that coordinate epigenetic regulation are themselves regulated by epigenetic modifications. Every regulatory RNA has to be transcribed: every highlighter, eraser or decoder protein has to be transcribed and translated. Just as for any other gene or protein, these processes are coordinated in part by transcription factors, DNA methylation, histone modification, chromatin remodelling, nuclear location and RNA, all working together.

(EPIGENET
IC) ▮ ▮ MODIFI
CATIONS, ▮ ▮
THEIR ▮ ▮ ▮
[INTERACTI
ONS] ▮ ▮ WITH
▮ ▮ ▮ EACH OT
HER ▮ & ▮ ▮ ▮ T
HEIR ▮▮▮▮ ▮ ▮
EFFECTS/ON
▮▮▮ (GENET
IC) ▮ ▮ ▮ REGU-
LATION ▮▮
▮ ▮ ▮▮▮ ...

... REPRESENT AN INCREDIBLY COMPLEX CODE THAT WILL KEEP SCIENTISTS BUSY FOR YEARS!

Epigenetics Explains What Genetics Alone Cannot

DNA methylation, histone modifications, chromatin remodelling and regulatory RNAs are involved in diverse processes throughout our lives, from the fertilized zygote's first few cell divisions to the creation of offspring and onwards into old age. The discovery and study of individual epigenetic modifications represent small pieces in the much larger puzzle of how these processes work.

Epigenetics is already starting to help explain cell differentiation, imprinting, the interplay between nature and nurture, and some of the other gaps in our knowledge of genetics.

IF YOU CAN'T EXPLAIN THE RESULTS OF AN EXPERIMENT, YOU CAN JUST SAY: "IT MUST BE EPIGENETICS".

The field isn't *quite* that advanced just yet, but some long-standing mysteries are indeed being solved.

Epigenetic Changes During Embryonic Development

The newly fertilized zygote inherits its chromosomes (and the proteins and regulatory RNAs attached to them) from both parents; it also receives some of the sperm cell's RNAs and proteins. However, most of its RNAs and proteins are derived from the mother's egg cell.

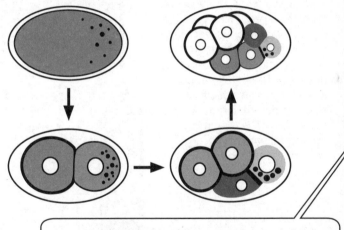

SOME MATERNAL PROTEINS AND RNAS ARE ARRANGED ASYMMETRICALLY WITHIN THE EGG CELL, AND AREN'T SHARED EQUALLY BETWEEN THE TWO CELLS THAT ARE PRODUCED DURING THE FIRST MITOSIS.

Because some of the molecules from the egg cell are involved in establishing epigenetic modification patterns, epigenetic differences between cells start to appear very early in embryonic development.

During the next few rounds of mitosis, in the first week after conception, all the cells of the early embryo undergo a drastic epigenetic reset. The overall amount of CpG methylation falls and then starts to climb again, a phenomenon called **epigenetic reprogramming***.

The DNA inherited from the egg cell is passively demethylated as cells divide (see page 48). In contrast, the paternal genome – which is tightly packaged with repressive protamine proteins, rather than histones (see page 56) – is actively demethylated, a much faster process (see page 49). These differences allow for parent-specific transcription of some genes, which is required for normal embryonic development.

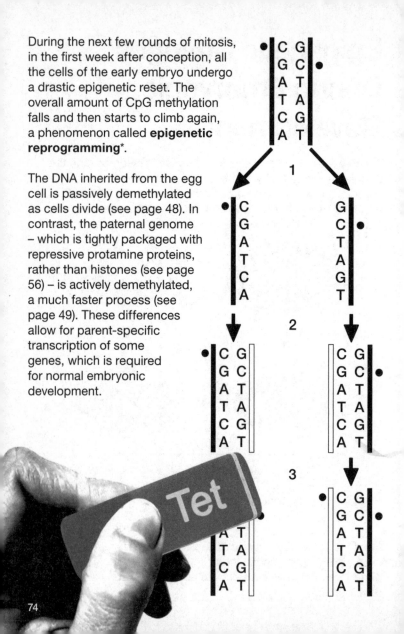

The remethylation of the maternal and paternal genomes coincides with a major wave of cell differentiation. The locations of the new methyl groups become increasingly divergent in different cells.

The initial divergence in epigenetic modification patterns stems from the first asymmetrical mitosis. The distinctive epigenetic landscape of each resulting cell directs the production of a unique combination of additional epigenetic regulators and transcription factors, thereby amplifying the cells' initial differences. Further rounds of this cycle help to define and maintain progressively distinct epigenetic patterns in each type of mature cell.

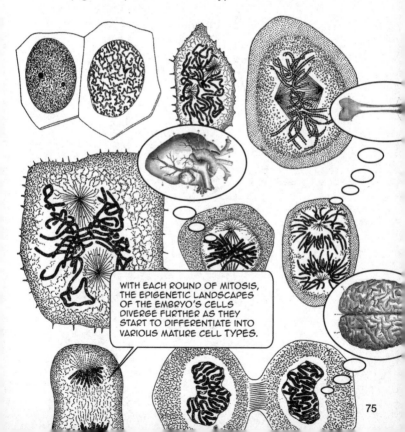

WITH EACH ROUND OF MITOSIS, THE EPIGENETIC LANDSCAPES OF THE EMBRYO'S CELLS DIVERGE FURTHER AS THEY START TO DIFFERENTIATE INTO VARIOUS MATURE CELL TYPES.

Although DNA methylation plays a critical role, other types of epigenetic regulation are also involved in driving cell differentiation during embryonic development. All of the RNA- and protein-based components of the epigenetic regulation network cooperate to fine-tune, spread and amplify the distinct epigenetic patterns that define each type of cell.

Once the cells reach full maturity, their epigenetic modification patterns (and therefore their gene transcription patterns, RNAs and proteins) stabilize and are copied through further rounds of mitosis. From this point on, the embryo focuses less on cell differentiation and increasingly on growth.

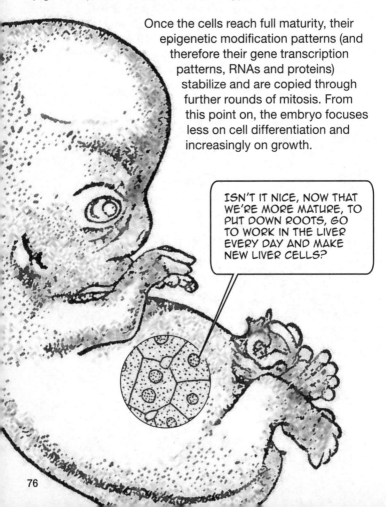

ISN'T IT NICE, NOW THAT WE'RE MORE MATURE, TO PUT DOWN ROOTS, GO TO WORK IN THE LIVER EVERY DAY AND MAKE NEW LIVER CELLS?

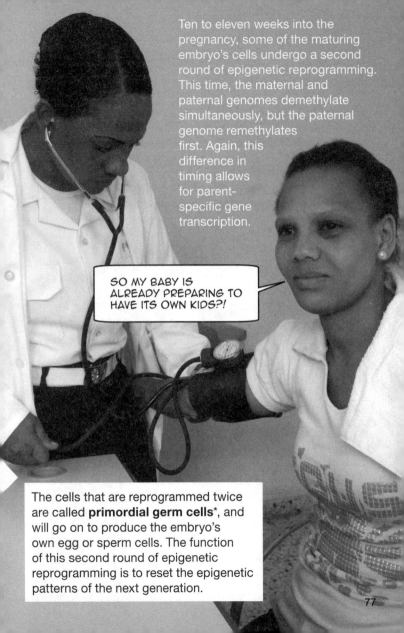

Ten to eleven weeks into the pregnancy, some of the maturing embryo's cells undergo a second round of epigenetic reprogramming. This time, the maternal and paternal genomes demethylate simultaneously, but the paternal genome remethylates first. Again, this difference in timing allows for parent-specific gene transcription.

SO MY BABY IS ALREADY PREPARING TO HAVE ITS OWN KIDS?!

The cells that are reprogrammed twice are called **primordial germ cells***, and will go on to produce the embryo's own egg or sperm cells. The function of this second round of epigenetic reprogramming is to reset the epigenetic patterns of the next generation.

Some parts of the genome escape one or both rounds of epigenetic reprogramming. For example, most repetitive DNA – which can cause harmful gene mutations when active – remains methylated during both rounds of reprogramming.

This ongoing repression of repetitive elements by piRNAs is important because the embryonic and germ cell development processes are so sensitive, complex and critical. The DNA replication errors and other mutations that repetitive elements can cause would be particularly dangerous during these times, potentially causing foetal abnormalities with serious life-long consequences, or even spontaneous miscarriage.

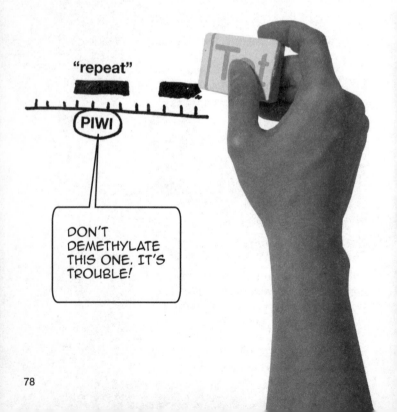

Many imprinted genes, which are transcribed from only one chromosome (see page 22), are involved in controlling cell growth and development. Imprinted genes therefore have to be tightly regulated in the embryo. Imprinted genes are associated with a piece of DNA called an **imprint control region (ICR)**.

Methylated ICRs are recognized and bound by a protein called ZFP57, which recruits partners that protect these regions from demethylation during the first epigenetic reset in early embryonic development. ZFP57 is not present in primordial germ cells. The unprotected imprinted genes therefore do demethylate and remethylate during primordial germ cell development.

The restricted transcription of imprinted genes from either the maternal or the paternal chromosome is crucial to their function. Mouse embryos that have been artificially given two copies of the maternal or paternal genome, rather than one copy of each, die at an extremely early stage of development, before they even implant into the uterus.

There are several human disorders caused by imprinting errors. Children with imprinting disorders are often born with intellectual disabilities and atypical physical characteristics, highlighting the importance of normal imprinting in embryonic development.

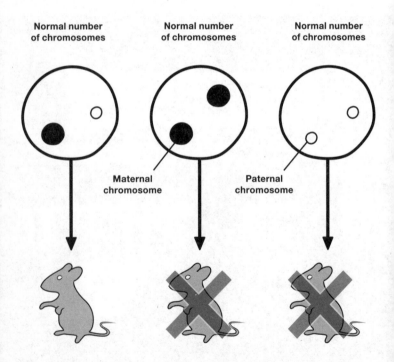

Normal number of chromosomes

Normal number of chromosomes

Normal number of chromosomes

Maternal chromosome

Paternal chromosome

Imprinted genes tend to cluster together in groups. Each group has its own imprint control region (ICR), which coordinates the regulation of the entire cluster via its own unique mechanism.

The methylation status of the ICR defines which genes in its cluster are transcribed from which chromosome. For instance, the ICR near the gene for a lncRNA called Kcnq1ot1 is methylated on the maternally-derived chromosome, but not on the paternal chromosome. Because DNA methylation represses transcription, Kcnq1ot1 is produced only from the paternal chromosome.

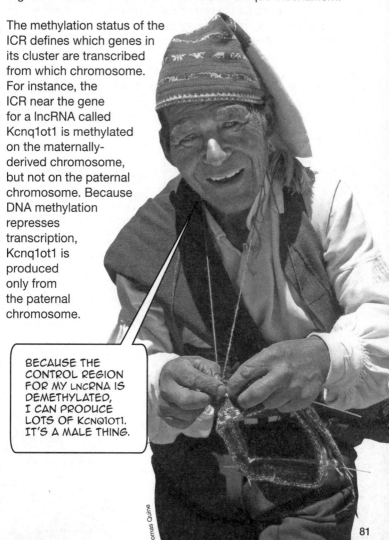

BECAUSE THE CONTROL REGION FOR MY LNCRNA IS DEMETHYLATED, I CAN PRODUCE LOTS OF KCNQ1OT1. IT'S A MALE THING.

The Kcnq1ot1 lncRNA is rather unadventurous and likes to stay close to home. One end of the strand binds to a complementary piece of DNA within the same cluster of imprinted genes, on the same chromosome from which it was transcribed. The other end recruits histone-modifying proteins that shut down the transcription of a nearby imprinted gene called Cdkn1c.

Because Kcnq1ot1 is only transcribed from the paternal chromosome, Cdkn1c is only produced from the maternal chromosome. In this way, maternal methylation of a single ICR defines the maternal transcription of one gene and the paternal transcription of another.

Some clusters of imprinted genes are controlled by regulatory RNAs and proteins that are only produced in certain cells; as a result, some genes are imprinted in some tissues, but not others.

ICR methylation patterns are reset in primordial germ cells (see page 77). Male embryos erase the maternal ICR methylation patterns they inherited from their mothers during this time, so that the sperm cells they will later produce will only contain chromosomes marked as paternal; female embryos do the reverse. Once set, ICR methylation patterns are maintained until the primordial germ cells of the next generation are formed.

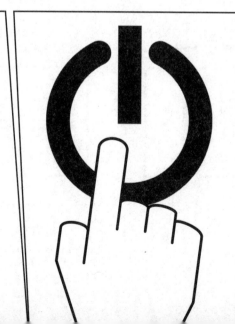

PRIMORDIAL GERM CELLS RESET THEIR IMPRINTED GENES SO THAT ALL CHROMOSOMES IN MATURE SPERM CELLS ARE MARKED "PATERNAL" AND ALL CHROMOSOMES IN MATURE EGG CELLS ARE MARKED "MATERNAL".

X Chromosome Inactivation

As we saw earlier, cells that contain two XX chromosomes inactivate one copy, to compensate for the different sizes of the X and Y chromosomes (see page 23). This phenomenon is also mediated by epigenetic mechanisms. In fact, X chromosome inactivation is the perfect demonstration of how different kinds of epigenetic regulation – DNA methylation, histone modifications and chromatin remodelling, all guided by a lncRNA – can cooperate to establish and maintain stable chromatin states.

In the earliest stages of XX embryo development, the paternally-inherited X chromosome is inactivated in every cell. However, this inactivation is reversed within a week of conception, during the first round of epigenetic reprogramming. Random inactivation of either the maternal or the paternal chromosome is established as the genome remethylates.

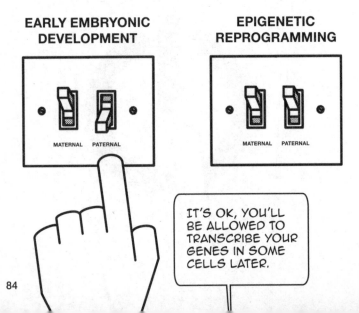

EARLY EMBRYONIC DEVELOPMENT

MATERNAL   PATERNAL

EPIGENETIC REPROGRAMMING

MATERNAL   PATERNAL

IT'S OK, YOU'LL BE ALLOWED TO TRANSCRIBE YOUR GENES IN SOME CELLS LATER.

Random X chromosome inactivation takes place independently in each cell of the early embryo. The same X chromosome remains inactivated in every mature cell that is descended from each of these first few cells. Because each cell's descendants tend to stay together in clusters, the organs and tissues of all XX female mammals come to resemble mosaics, with different X chromosomes inactivated in different regions.

Silenced X chromosomes are reactivated in female primordial germ cells, allowing random inactivation to repeat and form new mosaic patterns in the next generation.

THE PATCHES OF COLOUR ON THE COATS OF TORTOISESHELL AND CALICO CATS PROVIDE A VISIBLE EXAMPLE OF THE GENERAL PRINCIPLE THAT FEMALE MAMMALS ARE CHROMOSOMAL MOSAICS. EACH PATCH OF COLOUR REPRESENTS A CLUSTER OF SKIN CELLS DESCENDED FROM A SINGLE EARLY EMBRYONIC CELL THAT SHUT DOWN THE X CHROMOSOME CONTAINING THE GENE FOR EITHER ORANGE OR DARK FUR COLOUR.

How Our Environment Affects Our Genes

Epigenetics research is also revolutionizing the centuries-old question of "Nature or nurture?"

Certain chemicals can bind to specific **receptor proteins*** inside cells, or on the cell's outer surface. This triggers a **signalling cascade*** that passes the message on from protein to protein, and ultimately to the cell nucleus. Some cascades are initiated by molecules that originate outside the body (what we think of as "nurture"), and others by hormones and other chemicals that the body produces naturally ("nature"). The cell's ultimate response sometimes involves changes to the proteins and RNAs that regulate epigenetic modifications.

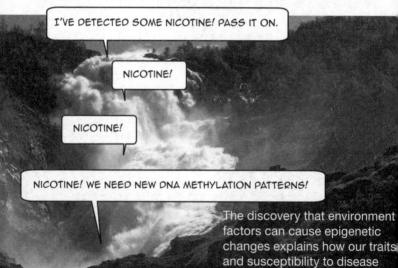

I'VE DETECTED SOME NICOTINE! PASS IT ON.

NICOTINE!

NICOTINE!

NICOTINE! WE NEED NEW DNA METHYLATION PATTERNS!

The discovery that environment factors can cause epigenetic changes explains how our traits and susceptibility to disease are affected by both nature *and* nurture. The boundaries between these "opposites" are blurring.

The epigenetic effects of the environment were first studied in an inbred mouse strain named after the agouti gene. Agouti mice have a piece of repetitive DNA close to their namesake gene that – when unmethylated – keeps the gene constantly activated, causing a yellow coat colour, obesity, type 2 diabetes and an increased risk of cancer. When this DNA is methylated, the agouti gene is silenced, causing darker, thinner and healthier mice.

When pregnant agouti mice eat methyl-rich supplements, such as folic acid, the repetitive DNA becomes more heavily methylated in the embryos' cells. The strength of the effect correlates with the extent of DNA methylation.

Scientists have exposed human cells to chemicals in the laboratory, and compared epigenetic modification patterns in people who've experienced exposures to the same chemical. These studies have identified many environmental factors that have epigenetic effects. Epigenetic modifiers range from the harmful – nicotine, benzene, arsenic, viral infections – to more benign molecules like folic acid and vitamin C.

Environmental exposures at any stage of life can affect our health. But just like agouti mice that experience lifelong consequences of their pregnant mother's diet, we're particularly vulnerable to epigenetic changes while we're developing – both in the womb and during childhood.

OUR DIETS, HABITS AND OTHER ENVIRONMENTAL FACTORS CAN CAUSE BOTH HARMFUL AND PROTECTIVE EPIGENETIC CHANGES IN OUR CELLS.

One example of how childhood environments can have long-term epigenetic effects involves children who experience physical or emotional abuse, who often go on to suffer from lifelong poor health. Even people with no conscious memory of the abuse carry an increased risk of heart disease, cancer, substance abuse, depression and other conditions. Abuse causes permanent DNA methylation changes. The initial trigger is thought to be the stress response hormone cortisol, which abused children produce in large amounts.

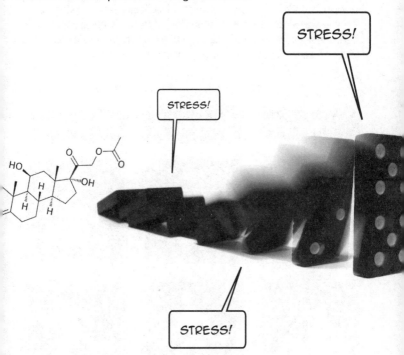

The hope is that research in this field will help to identify pharmaceutical or other interventions that can help to protect child abuse victims from developing health problems later in life.

Behaviour and environmental exposures during adulthood are important too. For example, we've known for centuries that exercise is beneficial to our health, but didn't know why. The benefits of burning more calories, and of increased cardiovascular fitness and muscle strength, were relatively easy to explain – but why would exercise also reduce the risk of cancer, dementia and depression?

At least part of the answer seems to involve exercise-induced changes in miRNA production and DNA methylation patterns. Exercise correlates with the silencing of genes that are involved in cell division and inflammation, which might help to explain the effects of exercise on cancer and other disease risks.

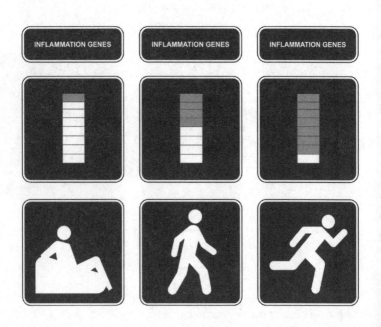

Not So Identical Twins

The ability of non-genetic factors to change epigenetic modification patterns helps to explain why identical twins aren't truly identical. Siblings who share the same DNA sequence, but who accumulate increasingly different experiences and environmental exposures as they age, provide a unique opportunity to study the impact of epigenetics on human health.

In 2005, Spanish geneticist **Manel Esteller** (b. 1968) compared the chromatin of pairs of identical twins at various ages, from birth to old age. As you'd expect, twins are very epigenetically similar at birth, but their DNA methylation and especially their histone modification patterns diverge gradually over time. Greater differences are seen between twins who have spent more time apart.

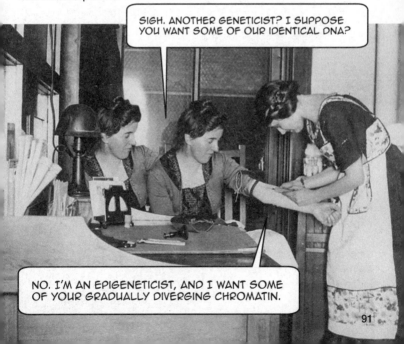

SIGH. ANOTHER GENETICIST? I SUPPOSE YOU WANT SOME OF OUR IDENTICAL DNA?

NO. I'M AN EPIGENETICIST, AND I WANT SOME OF YOUR GRADUALLY DIVERGING CHROMATIN.

Epigenetic twin studies range from the conventional, such as comparing cigarette smokers to their non-smoking twins, to the truly unique: NASA's 2015–16 experiment comparing identical twin astronauts, **Scott** and **Mark Kelly**, during and after Scott's year-long mission to the International Space Station. The results of NASA's cosmic epigenetics research are highly anticipated.

Back on Earth, in 2015 a Canadian team led by **Shiva Singh** found different DNA methylation patterns in blood cells from identical twin pairs in which one sibling has schizophrenia and the other does not. It's not yet known whether these differences are directly relevant to the disorder, but this type of research might eventually explain the non-heritable components of complex disorders – and possibly help to prevent disease in individuals with a high genetic risk.

Epigenetics research might also find applications in law enforcement. For example, epigenetic analysis could potentially solve crimes where DNA evidence indicates that one of two identical twins was to blame. This might sound like the plot of a Scandinavian crime drama, but it has actually happened several times, in countries around the world!

More broadly, beyond differentiating twins, epigenetic tests of crime scene evidence might eventually be able to indicate that the culprit was a cigarette smoker, heroin addict or exercise enthusiast. We don't know enough yet about the epigenetic effects of specific chemicals and behaviours, but the investigations (and crime dramas) of the future might just incorporate this kind of epigenetic profiling.

WELL, THE BISULPHITE SEQUENCING RESULTS ARE BACK. THERE'S A 92.7% CHANCE THAT THE CULPRIT SMOKES CIGARETTES AND EXERCISES A LOT.

Epigenetic Inheritance

Epigenetic reprogramming during early embryonic development, and again during the development of the primordial germ cells (see pages 74–7), is thought to act as a reset button which prevents the epigenetic changes that accumulate during an individual's lifetime from being passed on to the next generation.

In most cases, the epigenetic slate seems to be successfully wiped clean. However, as we'll see, there's evidence from some recent studies that some degree of epigenetic inheritance from parent to child can occur. This is a controversial area of research that could have wide-ranging implications – from nutrition during pregnancy to how we think about evolution!

THERE'S MOUNTING EVIDENCE THAT SOME EPIGENETIC CHANGES CAN PASS FROM PARENT TO CHILD – BUT THIS HAPPENS IN COMPLICATED AND SUBTLE WAYS.

One reason for the controversy is that it's difficult to distinguish genuine epigenetic inheritance from the effects of exposures in the womb or during early childhood.

Cells are particularly vulnerable to environmental exposures while they're undergoing epigenetic reprogramming. A pregnant woman's environment during the first week of embryonic development can create epigenetic changes with lifelong consequences for her child. Changes later in the pregnancy, around weeks ten to eleven, when the embryo's primordial germ cells are developing can also affect her future grandchildren.

But can the mother's prepregnancy environmental exposures – or the environment of the father – also affect the next generation?

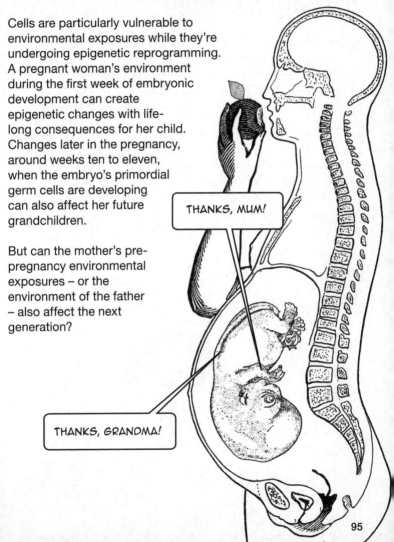

THANKS, MUM!

THANKS, GRANDMA!

95

Epigenetic Inheritance in Animal Models

As with many other complex scientific problems, the easiest way to study epigenetic inheritance is to perform very carefully controlled experiments using laboratory animals. Animal research (which is very strictly regulated, with independent expert approval needed for every study) can use techniques that would be impractical or prohibited in human studies. For example, the diets of pregnant and nursing mothers can be very strictly controlled; surrogate mothers can be used; and newborn animals can be placed with unrelated foster mothers. These approaches allow scientists to isolate and characterize the contributions of genetics, epigenetics, parenting styles and exposures during pregnancy, to gain a better understanding of how traits are inherited.

YOU EXERCISED DURING PREGNANCY, AND YOU DIDN'T? OK, TIME TO SWITCH YOUR BABIES.

COVER YOUR EARS, POPPET.

WHAT IS WRONG WITH THAT NASTY MAN?!

The methylation-dependent physical characteristics of agouti mice (see page 87) are very useful in epigenetic inheritance research.

Under normal circumstances, the degree of methylation of the piece of repetitive DNA close to the agouti gene varies randomly between individuals, and determines where each mouse falls on a spectrum from yellow, obese and diabetic (unmethylated repeat) to dark, slim and healthy (fully methylated repeat).

The methylation status of male agouti mice doesn't affect their offspring; whether they're yellow or black, they sire litters with a mix of coat colours and weights.

Unlike males, female agouti mice produce offspring more like themselves: darker mothers produce greater numbers of methylated, darker, slimmer offspring than do yellow mothers.

Because only the mother's DNA methylation status affects the next generation, the mechanism could theoretically involve exposures during pregnancy, rather than direct epigenetic inheritance. Perhaps obese, diabetic pregnant mothers provide more sugar, hormones or something else that affects the developing embryos directly.

To rule out this possibility, Australian geneticist **Emma Whitelaw** took newly fertilized zygotes from dark mothers and implanted them into the wombs of yellow mothers, and vice versa. The methylation status of the biological mother, not the surrogate mother, affected the next generation, providing evidence for the direct inheritance of epigenetic modifications.

Like agouti, the **axin gene** is controlled by a nearby piece of repetitive DNA in some mouse strains. Unmethylated repeats interfere with axin transcription, changing the sequence of the mRNA and the corresponding protein. The mutated axin protein introduces a kink into the mouse's tail.

Unlike agouti, both male and female mice can transmit their axin gene methylation status to the next generation: fathers and mothers with more heavily methylated repeats and straighter tails produce more straight-tailed offspring than do parents with kinked tails. Again, this finding provides evidence for direct epigenetic inheritance, independent of exposures during pregnancy.

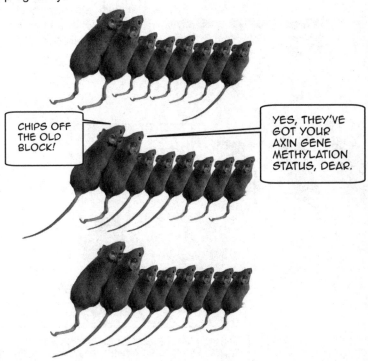

CHIPS OFF THE OLD BLOCK!

YES, THEY'VE GOT YOUR AXIN GENE METHYLATION STATUS, DEAR.

Additional evidence for direct epigenetic inheritance comes from studies by American neuroscientist **Yasmin Hurd**, in which adolescent rats were exposed to tetrahydrocannabinol (THC), the active molecule in cannabis.

After all the THC had cleared from the rats' bodies, Hurd's team allowed the test subjects to mate with untreated rats. The resulting pups were given to foster mothers with no previous THC exposure, in case the drug had affected the biological mothers' parenting skills.

When the pups matured, they were introduced to a system that required them to exert physical effort to self-administer heroin.

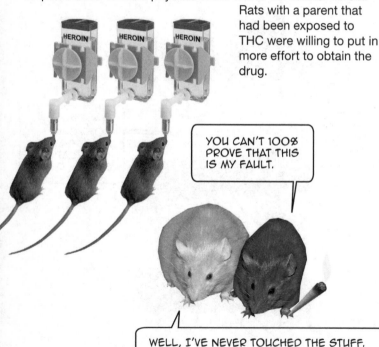

Rats with a parent that had been exposed to THC were willing to put in more effort to obtain the drug.

Both male and female rats that were given THC as adolescents produced offspring that were predisposed to heroin addiction. The next generation – the grandchildren of the treated rats – also displayed altered patterns of behaviour.

Yasmin Hurd's original 2014 study found that the offspring of THC-treated rats have abnormal patterns of mRNA and protein production in their brain cells. Some of the affected proteins act as receptors for cannabis-based drugs, and are involved in compulsive behaviours and addiction. A follow-up study by the same team in 2015 found brain-specific DNA methylation changes that correlate with the altered patterns of gene transcription.

101

Epigenetic modifications can also be inherited indirectly. For instance, some mother rats lick their pups often, while others are less attentive. Being licked causes the demethylation of genes that are involved in responding appropriately to stressful situations; unlicked, neglected pups grow into more stressed adults. Cross-fostering experiments show that the pups' stress levels are established by the parenting skills of the foster mother, rather than the biological mother.

Some of the excess DNA methylation in neglected rats affects genes that are involved in parenting. The stressed rats therefore don't lick their own pups, perpetuating the cycle of neglect even though the epigenetic modifications aren't passed on directly.

IT STRESSES ME OUT WHEN YOU DON'T LICK ME!

BLAME YOUR GRANDMA! YOU'LL UNDERSTAND WHEN YOU HAVE PUPS OF YOUR OWN.

Human Epigenetic Inheritance:
The Dutch Hunger Winter

During the harsh winter of 1944–5, the Nazis blocked all food imports to the Netherlands, causing a devastating famine. It's estimated that 20,000 people died of starvation before food supplies returned to normal with the liberation of the Netherlands in May 1945.

The sharp boundaries of the period of starvation, and the survivors' subsequent access to a comprehensive public healthcare system, presented a unique research opportunity for epigenetic inheritance studies. Teams of Dutch and international scientists have been studying the survivors and their descendants ever since the famine ended.

AND THIS
HAS TO
LAST US
ALL WEEK ...

Dutch-born epidemiologist **Bertie Lumey** has studied survivors of the Dutch Hunger Winter, and found that people conceived during the famine (whose mothers were starving during early pregnancy) have an increased risk of obesity, diabetes and heart disease. Lumey's team showed in 2008 that these risks correlate with reduced DNA methylation levels in imprinted genes that are involved in metabolism.

Children whose mothers first experienced starvation later in pregnancy weren't affected in the same way, suggesting that earlier stage embryos are more vulnerable to epigenetic changes caused by the environment. However, this latter group of people did have slightly lower-than-average birth weights and lower-than-average obesity rates over the course of their lifetimes.

The first statistical analysis of data from the third generation of Dutch Hunger Winter survivors suggested that the health effects of the famine persist in the survivors' grandchildren. However, a later study found no evidence for a higher risk of specific diseases, although it did find that third-generation individuals generally have more body fat and worse overall health. It's important to note that the early research findings were well publicized, and survivors' grandchildren might have changed their lifestyles to mitigate their higher disease risk before the later study was completed.

These rather confusing results highlight the difficulty of studying human populations in general, and maternal transmission in particular. Even if the famine really does still affect the third generation, the mechanism might involve exposure of the second generation's developing primordial germ cells to starvation in the womb, rather than genuine epigenetic inheritance.

SURVIVOR OF THE DUTCH HUNGER WINTER

DAUGHTER
Ill health as a result of the famine

GRANDDAUGHTER
Ill health – not conclusively via epigenetic inheritance

Human Epigenetic Inheritance: Överkalix

A Swedish team led by **Michael Sjöström** and **Lars Olov Bygren** have found stronger evidence for human epigenetic inheritance, thanks to meticulous record keeping in the northern town of Överkalix. Combining the town's harvest and birth records from 1890 onwards allowed the researchers to identify people who experienced either feast or famine at various stages of their lives, and to trace the medical histories of their descendants.

One of the key findings of the Överkalix study concerns sex-specific patterns of disease risk in later generations: women who were in their mother's womb during a period of famine passed on a higher risk of cardiovascular disease to their granddaughters, but not to their grandsons.

THIS FAMINE COULD AFFECT THE WOMEN IN MY FAMILY FOR GENERATIONS.

The Överkalix researchers also found sex-specific patterns of inheritance down the male line: men who had experienced a feast year between the ages of nine and twelve had grandsons – but not granddaughters – with shorter-than-average lifespans. Conversely, a period of famine in the same pre-puberty years resulted in healthier grandsons with longer lifespans.

Sperm precursor cells are differentiating and maturing in the years just before puberty, and thus might be vulnerable to epigenetic changes induced by the environment. The sex-specific patterns of disease risk suggest that imprinted genes might be involved, although no clear mechanism has yet been found.

Boys seem to be particularly vulnerable to other environmental exposures during the years just before puberty. A British study led by geneticist **Marcus Pembrey** found that men who smoked cigarettes before puberty generally went on to have sons – but not daughters – with more body fat, regardless of whether the men were still smokers when their children were conceived.

A Taiwanese and British study led by **Barbara Boucher** found a similar association between betel nut chewing in pre-adolescent boys and a higher risk of metabolic syndrome (a precursor to diabetes and cardiovascular disease) in their future offspring, although this latter study did not find sex-specific differences.

SMOKING MAY AFFECT YOUR SPERM CELLS AND HARM YOUR FUTURE CHILDREN AND GRANDCHILDREN.

Mechanisms of Epigenetic Inheritance

Animal experiments have provided credible evidence for genuine epigenetic inheritance; human population studies have hinted that the same phenomenon might exist in our own species. The findings thus far have attracted a lot of attention and interest from scientists and non-scientists alike. However, some human studies have been over-simplified and over-interpreted, especially by the media, and many scientists remain sceptical about human epigenetic inheritance.

The sense of caution that surrounds this field is heightened further by the existence of many unanswered questions, including those about the actual mechanism of inheritance. Researchers have found correlations between certain traits that are thought to be affected by epigenetic inheritance and DNA methylation changes in specific genes, but this doesn't necessarily mean that the two phenomena are directly related.

SO I CAN'T BLAME MY GRANDFATHER FOR THIS?

NOT JUST YET. NOT UNLESS YOU'RE A MOUSE.

Coming back to epigeneticists' favourite mouse, the inheritance of maternal agouti gene methylation status (see pages 98–9) initially seemed like it should be relatively easy to explain. Most repetitive DNA doesn't undergo epigenetic reprogramming (see page 78); therefore, the mother's methylation status might simply persist in the embryo.

However, Emma Whitelaw's team found in 2006 that the repeat near the agouti gene is completely demethylated and then remethylated during early embryonic development. Like the rest of the genome, the maternal and paternal copies are demethylated at different speeds, by different means. This might explain why only female mice can pass on their methylation status, but the mechanism remains a mystery.

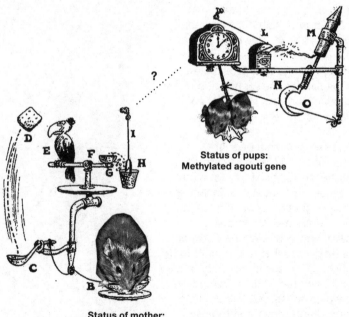

Status of pups:
Methylated agouti gene

Status of mother:
Methylated agouti gene

In 2009–10, British and Swiss teams led by **David Miller** and **Antoine Peters**, respectively, found some interesting patterns in human sperm chromatin.

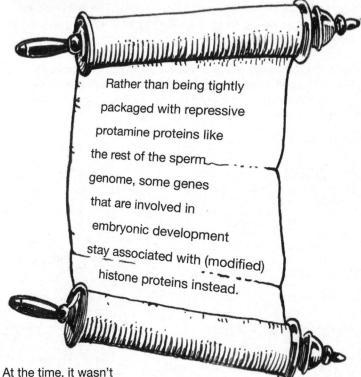

Rather than being tightly packaged with repressive protamine proteins like the rest of the sperm genome, some genes that are involved in embryonic development stay associated with (modified) histone proteins instead.

At the time, it wasn't thought that histone modifications could be inherited. However, in 2014 **Susan Strome**'s work, which showed histone modifications being passed to newly-forming strands of DNA (see page 57), also found that histone methylation can be inherited by the next generation. Strome's study was performed in worms, and it's not yet known whether the same mechanism exists in humans. However, it's certainly possible that histones are involved in transmitting some forms of epigenetic information between generations.

RNA molecules might also mediate epigenetic inheritance in some cases. Zygotes inherit the RNAs attached to the maternal and paternal chromosomes, plus some of the sperm cell's free-floating RNAs and all of the egg cell's RNAs. These molecules might be responsible for re-establishing parental epigenetic modification patterns after reprogramming.

Swiss neuroscientist **Isabelle Mansuy** and Australian biologist **Michelle Lane** have demonstrated that traumatic stress and obesity, respectively, can alter the repertoire of small RNAs in mouse sperm cells.

THE DAUGHTERS OF FORMERLY OBESE MALE MICE THAT HAD EXERCISED HAD LESS BODY FAT AND HEALTHIER INSULIN RESPONSES THAN THE DAUGHTERS OF SEDENTARY OBESE MICE.

In 2016, a Danish team led by **Romain Barrès** published evidence that RNA-based mechanisms are also involved in human inheritance. Sperm cells from obese men have distinct small RNAs and DNA methylation patterns, some of which change following surgery-induced weight loss. Genes that are thought to play a role in controlling appetite are among those that undergo methylation changes.

Epigenetics in Evolution

The discovery of epigenetic inheritance has raised an interesting question about evolution: if some of the epigenetic modifications that we acquire during our lifetimes can be passed on to our children and grandchildren, can they also affect how species evolve over many hundreds of generations?

The idea that acquired characteristics can be inherited, helping to shape how species change over time, isn't new. Today, it's most closely associated with French biologist **Jean-Baptiste Lamarck** (1744–1829), and is often referred to as "Lamarckism", but the concept itself is much older. Lamarck did, however, formalize the idea of this "soft inheritance" in 1809.

'ALL THE ACQUISITIONS OR LOSSES WROUGHT BY NATURE ON INDIVIDUALS ... ARE PRESERVED BY REPRODUCTION TO THE NEW INDIVIDUALS WHICH ARISE.'

Although Lamarckism and related theories were popular for a long time, they were not without flaws. For example, despite generations of pedigree dogs having their ears and tails docked, their puppies clearly keep being born with long ears and tails.

In 1859, British naturalist **Charles Darwin** (1809–82) published his own theory of evolution by natural selection.

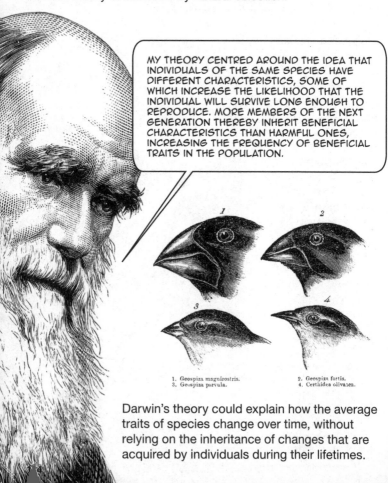

MY THEORY CENTRED AROUND THE IDEA THAT INDIVIDUALS OF THE SAME SPECIES HAVE DIFFERENT CHARACTERISTICS, SOME OF WHICH INCREASE THE LIKELIHOOD THAT THE INDIVIDUAL WILL SURVIVE LONG ENOUGH TO REPRODUCE. MORE MEMBERS OF THE NEXT GENERATION THEREBY INHERIT BENEFICIAL CHARACTERISTICS THAN HARMFUL ONES, INCREASING THE FREQUENCY OF BENEFICIAL TRAITS IN THE POPULATION.

1. Geospiza magnirostris.
2. Geospiza fortis.
3. Geospiza parvula.
4. Certhidea olivasea.

Darwin's theory could explain how the average traits of species change over time, without relying on the inheritance of changes that are acquired by individuals during their lifetimes.

After Mendel's work on the laws of inheritance (page 25) was rediscovered, and as we started to learn more about genes, chromosomes, DNA and mutations, it became obvious that only Darwin's theory of evolution was consistent with our understanding of genetics. The idea of soft, or Lamarckian, inheritance was essentially discredited.

The discovery of epigenetic inheritance in the late 1990s seemed to throw a lifeline to the older theory, however. Environmental exposures do sometimes seem to pass to later generations. Could there be room for a version of Lamarckism as one component of modern evolutionary theory, after all?

SOFT INHERITANCE THEORIES OF EVOLUTION HAVE BEEN OUT OF FAVOUR FOR DECADES.

BUT COULD AN UPDATED THEORY OF EVOLUTION MAKE ROOM FOR BOTH OUR IDEAS?

The modern understanding of Darwinian evolution focuses on inherited DNA sequence changes that affect individuals' characteristics and behaviour. However, the discovery of epigenetic modifications has proven that there's more to life than the DNA sequence alone: epigenetic modifications help to define gene activation patterns, and stable changes to these patterns can also be very important in evolution. For example, earlier activation or later silencing of a gene that promotes neuron growth during embryonic development might help to increase intelligence.

THE DNA SEQUENCES OF THE HUMAN AND CHIMPANZEE GENES THAT PROMOTE HAIR GROWTH ARE EXTREMELY SIMILAR, BUT ARE ACTIVATED IN MANY MORE PARTS OF A CHIMP'S BODY.

A 2013 study led by **Andrew Sharp** and **Tomas Marques-Bonet** compared DNA methylation patterns in humans, chimpanzees, bonobos, gorillas and orangutans. One hundred and seventy genes with human-specific methylation patterns were identified; some of these genes are known to have functions in the brain, an organ that's of particular interest in human evolution.

Some of the 170 genes with human-specific methylation patterns code for proteins that are identical to their ape counterparts. This discovery reinforces the idea that changes to when and where genes are activated can be as important as changes to their sequence and function.

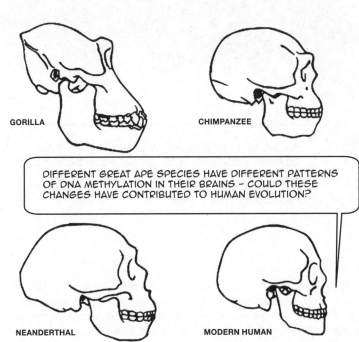

GORILLA

CHIMPANZEE

DIFFERENT GREAT APE SPECIES HAVE DIFFERENT PATTERNS OF DNA METHYLATION IN THEIR BRAINS – COULD THESE CHANGES HAVE CONTRIBUTED TO HUMAN EVOLUTION?

NEANDERTHAL

MODERN HUMAN

Despite the relationship between epigenetic modifications and gene activation patterns, and the role of altered gene activation patterns in evolution, Lamarckian evolution still doesn't really fit into modern evolutionary thinking. The reason is that evolution operates over many thousands of years, so the changes that drive it need to be very stable.

EPIGENETIC INHERITANCE LASTS FOR A FEW GENERATIONS AT MOST, AND IS REVERSIBLE: EVEN THE DARKEST AGOUTI MICE, WITH FULLY METHYLATED DNA, PRODUCE SOME YELLOW OFFSPRING WITH UNMETHYLATED DNA.

Epigenetic modifications might affect how individuals respond to environmental changes in the short term, but it seems very unlikely that long-term epigenetic inheritance can permanently change the characteristics of a species over hundreds and thousands of generations.

If epigenetic changes aren't stable enough to be directly inherited through multiple generations, how do inter-species epigenetic differences evolve?

The answer is that epigenetic evolution is driven by changes to DNA sequences: specifically, sequences that code for the RNAs and proteins that control epigenetic modifications, or sequences that are themselves modified. Any change to a DNA sequence that is involved in epigenetic regulation can potentially affect its function. Some changes will be beneficial; others will be harmful. Epigenetic evolution therefore occurs in a way that's fully consistent with the modern definition of Darwinism.

DIFFERENT DNA SEQUENCES CAUSE DIFFERENT EPIGENETIC MODIFICATION PATTERNS, WHICH CONTRIBUTE TO DIFFERENT PHYSICAL TRAITS ...

... WHICH CAN BE BENEFICIAL OR HARMFUL, LIKE ANY OTHER EVOLVABLE TRAITS.

Because there are such long gaps between human generations, a lot of evolution research focuses on species that reproduce (and therefore evolve) more rapidly, such as bacteria and worms.

Bacteria typically have several DNA methyltransferase proteins, each of which recognizes and methylates a specific DNA sequence. Different kinds of bacteria have evolved methyltransferase genes with different DNA sequences; these changes affect the target site preferences of the corresponding proteins, and therefore the patterns of DNA methylation in the genome.

Some species of nematode worms have recently completely lost one of their DNA methyltransferase genes. They therefore have different patterns of genome methylation compared to even their closest relatives.

In both cases, different DNA methylation patterns affect the amount, timing and location of gene activation.

HIM? OH, THAT'S MY COUSIN. THAT BRANCH OF THE FAMILY HASN'T BEEN QUITE THE SAME SINCE THEY DITCHED ONE OF THEIR METHYLTRANSFERASE GENES.

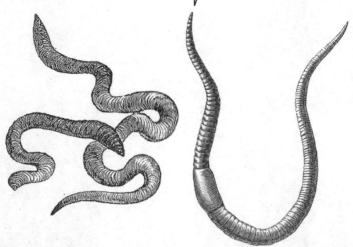

Regulatory RNAs (see pages 63–8) also play a role in epigenetic evolution. Changes in the sequence of an RNA strand can allow it to bind to different complementary DNA sequences, targeting entirely new genes, and/or can allow it to recruit different epigenetic modifiers to its existing target genes, changing their activation status.

WE'VE NEVER HAD ONE OF THESE DELIVERED TO THIS ADDRESS BEFORE!

lncRNA

Even minor RNA sequence changes can have significant effects on gene activation patterns. For this reason, mutations that affect regulatory RNAs might be just as important as those that alter the sequences of proteins – or maybe even more important. While some lncRNAs are conserved across multiple species, many others are unique to a single species. This evidence further supports the hypothesis that RNA evolution is involved in species evolution.

Epigenetic modifications can also help to shape the mutation and evolution of the DNA sequence itself. C bases are sometimes mistakenly converted into Ts during active DNA demethylation. Any such mistakes that occur in germ cells and that aren't corrected will be inherited by the next generation. Methylation target sites can thus be lost, affecting the transcription of nearby genes. There are hints that methylation patterns might affect the frequency of other types of mutation, too.

Although the specific consequences of these kinds of changes aren't well understood just yet, any DNA sequence change has the potential to cause new physical traits that contribute to how the species evolves.

Evolutionary theory has come a long way since the days of Lamarck and Darwin. We still don't know exactly how much epigenetics has contributed to the evolution of life on Earth, but we do now have a solid understanding of the underlying principles by which both DNA sequences and epigenetic modification patterns can evolve over time – and we've started to find examples of epigenetic changes that are consistent with these principles and with other aspects of modern Darwinism.

However, epigenetics is still a very young field, and there are probably more surprises ahead. Luckily, theories about evolution can themselves evolve!

Epigenetics in Disease: Ageing

When everything works as it should, epigenetic modifications help to establish and maintain the patterns of gene activation and protein production that are necessary for normal embryonic development and for the continued functions of our cells throughout our lifespans.

However, our cells and bodies are incredibly complicated systems. As such, there's a lot of potential for molecule, cell, tissue and organ functions to be perturbed at any stage of life, but especially as we age. Small initial changes can ripple out to cause serious problems, resulting in general poor health or in specific types of disease.

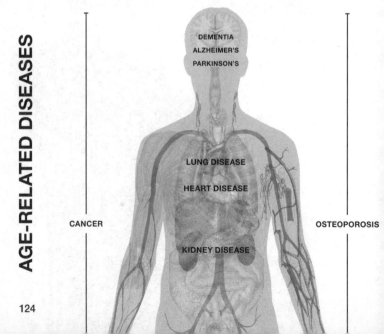

AGE-RELATED DISEASES

DEMENTIA
ALZHEIMER'S
PARKINSON'S

LUNG DISEASE

HEART DISEASE

CANCER

OSTEOPOROSIS

KIDNEY DISEASE

Epigenetic modification patterns can change after exposure to certain environmental factors (see pages 87–91). They're also subject to gradual, *random* changes over time. This latter phenomenon is known as **epigenetic drift** and is thought to play an important role in ageing.

Epigenetic drift causes different changes in each cell and each individual, but follows predictable general patterns. A few genes accumulate higher levels of DNA methylation over time, but the overall pattern is that the total amount of DNA methylation slowly diminishes as we age. Studies in mice have shown that this demethylation results in the gradual reactivation of silenced genes; these types of changes can alter the behaviour of cells in harmful ways.

IT'S JUST A CASE OF EPIGENETIC DRIFT – IT SHOULD FOLLOW A PREDICTABLE PATTERN.

Epigenetic drift can cause the partial differentiation of stem cells. This irreversible process reduces the number of active stem cells in the body that are available to replace dying mature cells. Heart muscle, skin elasticity and other tissues and functions can deteriorate as a result.

Epigenetic changes also occur in cells that are affected by specific diseases of ageing, from cancer to Alzheimer's and Parkinson's diseases, osteoporosis and heart failure. Given the complexity of epigenetic regulation, though, it's important not to over-interpret the evidence. Not every epigenetic difference between healthy and abnormal cells is meaningful, and some differences might be a response to the disease rather than a cause.

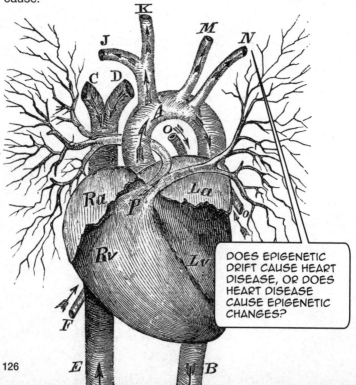

Epigenetics in Disease: Inherited Mutations in Epigenetic Regulators

Many human genetic disorders are caused by the mutation of a single gene. Mutations can be passed down from one or both parents, or can arise spontaneously in the egg, sperm or zygote. Familiar examples include sickle cell anaemia (caused by mutation of the haemoglobin protein that transports oxygen around the body) and cystic fibrosis (caused by mutation of a protein that pumps salt across cell membranes).

In the same way, some rare genetic disorders are caused by mutations in genes that play a role in epigenetic regulation. In contrast to diseases of ageing, these genetic disorders have an obvious epigenetic cause.

WHY ISN'T THE SICKLE CELL MUTATION REMOVED BY NATURAL SELECTION?

BECAUSE ONLY PEOPLE WHO INHERIT TWO FAULTY GENES ACTUALLY GET SICKLE CELL ANAEMIA. IF YOU ONLY INHERIT ONE MUTATED COPY, IT GIVES YOU SOME IMMUNITY TO MALARIA, MAKING IT A BENEFICIAL MUTATION THAT ACTUALLY *INCREASES* YOUR CHANCE OF SURVIVAL.

Kabuki syndrome is an example of a genetic disorder caused by a mutation in an epigenetic regulator. Fewer than 1 in 30,000 children are affected by this disorder, which causes distinctive atypical facial and skeletal features, among other symptoms.

About three quarters of cases are caused by inherited mutations in the MLL2 gene. The MLL2 highlighter protein usually adds transcription-activating methyl groups to histone tails; mutation of the protein, as occurs in those with Kabuki syndrome, eliminates this function.

Other cases of Kabuki syndrome involve mutations in the KDM6A gene, which normally codes for an eraser protein that removes repressive methyl groups from histones. The effects of KDM6A and MLL2 mutations are essentially the same, both involving the inappropriate repression of gene transcription.

IT OWES ITS NAME TO A TRADITIONAL FORM OF JAPANESE THEATRE, KABUKI, THE STAGE MAKEUP OF WHICH WAS THOUGHT TO RESEMBLE THE FACIAL CHARACTERISTICS OF THOSE WITH THE DISORDER.

Rett syndrome is a genetic disorder caused by spontaneous mutations in the MECP2 gene. "MECP2" stands for methyl-CpG-binding protein 2; as the name suggests, this decoder protein's normal function is to bind to methylated C bases, an essential step towards transcriptional repression. The effect of the mutation is that babies with Rett syndrome develop normally until they're around six months old, at which point their growth and development become atypical.

Unlike Kabuki syndrome, which affects both boys and girls, Rett syndrome is only ever seen in girls. Estimates of the disorder's prevalence vary from about 1 in 10,000 to about 1 in 20,000.

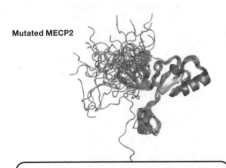

Mutated MECP2

THE MUTATION THAT CAUSES RETT SYNDROME BREAKS THE CONNECTION BETWEEN DNA METHYLATION AND TRANSCRIPTIONAL REPRESSION; GENES THAT ARE SUPPOSED TO BE SILENCED ARE ACTIVATED INSTEAD.

T

C

G

A

A

A

G

C

G

T

A

C

MECP2, the gene which mutates to cause Rett syndrome, is an X chromosome gene, so XY males only receive one copy. A mutation in this single copy causes spontaneous miscarriage of XY embryos in early pregnancy.

XX females inherit two MECP2 genes. Embryos that receive one mutated and one normal copy of the gene can survive to full term, but the babies go on to develop Rett syndrome. Random X chromosome inactivation causes a mosaic effect in which some XX cells within the same body only use the normal copy of the MECP2 gene, and others only use the mutated copy. The exact symptoms of Rett syndrome therefore vary between affected individuals, depending on the patterns of chromosome inactivation in the different tissues of their bodies.

	Normal MECP2 Gene on 1st X Chromosome	Mutated MECP2 Gene on 2nd X Chromosome	Effect
Brain Cell		Inactivated	Fewer neurological symptoms
Brain Cell	Inactivated		More neurological symptoms
Muscle Cell		Inactivated	More coordinated movement
Muscle Cell	Inactivated		Less coordinated movement

Epigenetics in Disease: Imprinting Errors

Genetic disorders can also be caused by errors that affect imprinting – the transcription of certain genes from only the maternal or only the paternal chromosome (see pages 79–83). Imprinting disorders can result from the mutation or loss of the imprint control region (ICR) that controls each cluster of imprinted genes, or of the RNAs and proteins that help to establish the methylation status of the ICR.

Imprinting errors effectively create two maternal or two paternal copies of the associated cluster of imprinted genes. This can cause either a double dose, or the complete loss of transcription, of imprinted genes. The exact effects of these gene dosage changes depend on the functions of the genes involved, but can seriously affect the ways in which cells and organs develop and function.

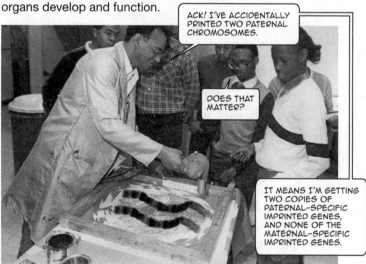

ACK! I'VE ACCIDENTALLY PRINTED TWO PATERNAL CHROMOSOMES.

DOES THAT MATTER?

IT MEANS I'M GETTING TWO COPIES OF PATERNAL-SPECIFIC IMPRINTED GENES, AND NONE OF THE MATERNAL-SPECIFIC IMPRINTED GENES.

Beckwith-Wiedemann syndrome is an example of an imprinting disorder. It's associated with the disruption of a cluster of imprinted genes on chromosome 11. The exact symptoms of the syndrome depend on the type of mutation inherited by each affected individual. Some ICR mutations completely abolish the transcription of an imprinted gene that helps to inhibit mitosis; others create a double dose of a growth promoting protein. As such, Beckwith-Wiedemann syndrome is usually characterized by symptoms such as rapid growth during childhood and an increased likelihood of developing childhood cancer.

Similar imprinting errors are sometimes acquired by mature cells, which can contribute to the development of cancer in adults.

NOT EVERY CHILD WITH BECKWITH-WIEDEMANN SYNDROME WILL DEVELOP CANCER – BUT THESE IMPRINTING ERRORS DO INCREASE THEIR RISK LEVELS.

The Epigenetics of Cancer

Cancer, a group of diseases characterized by uncontrolled cell growth and division, is the leading cause of death worldwide. About half of us will be diagnosed with some form of the disease at some point in our lives.

GLOBALLY, THERE WERE MORE THAN 14 MILLION NEW CASES AND 8 MILLION DEATHS FROM CANCER IN 2012.

THE NUMBERS ARE EXPECTED TO RISE IN FUTURE YEARS.

Every cancer diagnosis affects not only the individual patient but their loved ones too, adding up to an enormous global burden. The disease is also immensely expensive, with over 100 billion US dollars spent globally each year on cancer drugs alone.

There are many different causes of cancer, including inherited gene mutations in about 5 to 10 per cent of cases, but exposure to chemicals and other environmental factors also plays a huge role.

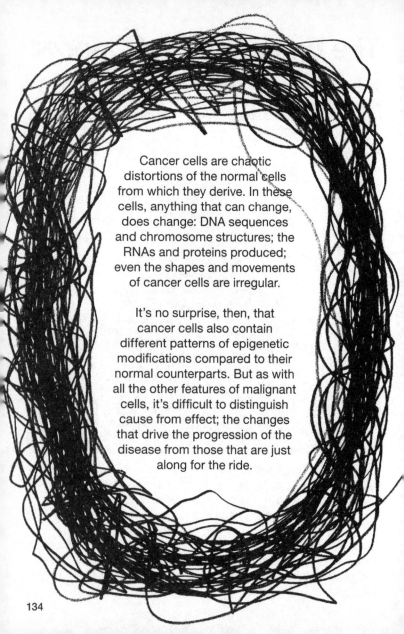

Cancer cells are chaotic
distortions of the normal cells
from which they derive. In these
cells, anything that can change,
does change: DNA sequences
and chromosome structures; the
RNAs and proteins produced;
even the shapes and movements
of cancer cells are irregular.

It's no surprise, then, that
cancer cells also contain
different patterns of epigenetic
modifications compared to their
normal counterparts. But as with
all the other features of malignant
cells, it's difficult to distinguish
cause from effect; the changes
that drive the progression of the
disease from those that are just
along for the ride.

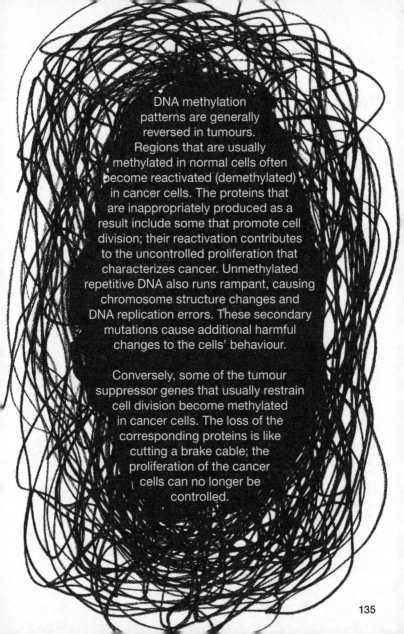

DNA methylation patterns are generally reversed in tumours. Regions that are usually methylated in normal cells often become reactivated (demethylated) in cancer cells. The proteins that are inappropriately produced as a result include some that promote cell division; their reactivation contributes to the uncontrolled proliferation that characterizes cancer. Unmethylated repetitive DNA also runs rampant, causing chromosome structure changes and DNA replication errors. These secondary mutations cause additional harmful changes to the cells' behaviour.

Conversely, some of the tumour suppressor genes that usually restrain cell division become methylated in cancer cells. The loss of the corresponding proteins is like cutting a brake cable; the proliferation of the cancer cells can no longer be controlled.

Some of the acquired DNA methylation changes in cancer cells affect imprint control regions that control the chromosome-specific transcription of imprinted genes.

Inappropriate ICR methylation mimics certain inherited imprinting disorder mutations, such as those that cause Beckwith-Wiedemann syndrome (see page 132), in that it effectively creates two maternal or two paternal copies of the associated cluster of imprinted genes. Many imprinted genes are involved in embryonic development, and therefore have functions that affect cell growth or division. Upsetting the ratios of the corresponding RNAs and proteins can cause excess cell proliferation, causing or contributing to the progression of cancer.

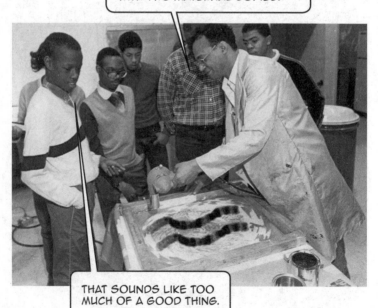

CANCER HAS AFFECTED THE ICR ON THIS GENE – WE'LL END UP WITH TWO MATERNAL COPIES.

THAT SOUNDS LIKE TOO MUCH OF A GOOD THING.

Like DNA methylation patterns, the regulatory RNAs produced by cancer cells can be radically different from those of the corresponding normal cells.

These changes can have particularly serious consequences. A single miRNA or circular RNA can affect the translation of many different proteins (see page 66). Abnormal production of lncRNAs and piRNAs can also have profound effects; because they attach to specific DNA sequences, these RNAs can initiate the first step in the complex process that specifies which epigenetic modifications should be made to which genes (see pages 64–5). Changes to this first step can easily ripple out to affect the transcriptional regulation of large parts of the genome.

DNA methylation changes
Histone modification changes
Chromatin structure changes
Cell behaviour changes
Cancer

Cancer cells contain altered histone modification patterns, which reinforce the gene activation changes initiated by the new pattern of DNA methylation, and vice versa.

In 2008, American geneticist **Kevin White** was the first to report changes in how and when specialized histone variant proteins (see page 56) replace the standard histones in malignant cells. The insertion of certain histone variants into nucleosomes at inappropriate locations and times can affect the transcription of nearby genes; the failure to insert other variants at the usual time and place can prevent DNA repair, accelerating the chaos within the cancer cell.

Finally, even the locations of some genes within the nucleus (see page 62) are different in cancer cells!

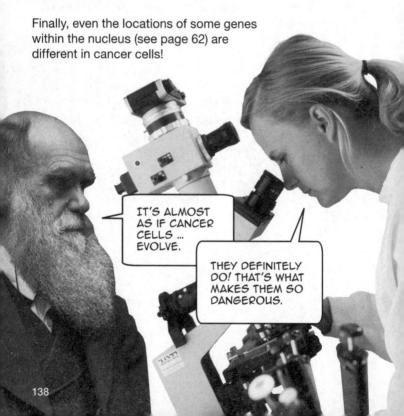

IT'S ALMOST AS IF CANCER CELLS ... EVOLVE.

THEY DEFINITELY DO! THAT'S WHAT MAKES THEM SO DANGEROUS.

Are the abnormal epigenetics of cancer cells the cause, or just an effect, of the disease?

Cancer usually begins with a DNA sequence change. There are many possible causes: inherited gene mutations, ultraviolet light, hepatitis or human papillomavirus infections, certain chemicals such as asbestos, or random DNA replication errors. Additional mutations accumulate as the cell becomes increasingly abnormal.

In some cases, the epigenetic changes seen in cancer cells are mere collateral damage – a response to earlier DNA mutations in the same cell. In others, epigenetic changes actively contribute to the chaos, for example by reactivating a silenced gene to make an already abnormal cell even more malignant. More rarely, epigenetic changes are responsible for initiating the whole process.

Some cancer-initiating epigenetic changes are probably the result of random chance – drift (see page 125) – or an error during mitosis.

Some environmental exposures can also initiate cancers via epigenetic changes. For example, DNA methylation patterns change after exposure to well-known cancer-causing agents such as tobacco smoke, and to new threats like Bisphenol A (BPA), a component of some plastics that has been found to leach out of drinking bottles and into the body.

However, tobacco smoke also damages the DNA directly, so it's hard to determine how much of its effect is exerted via epigenetics. In general our environments are so complex, containing thousands of harmful and beneficial factors, that it's extremely difficult to identify some of the subtler effects.

IT'S OK. I EXERCISED EARLIER. ALL THESE POSITIVE AND NEGATIVE EPIGENETIC EFFECTS CANCEL EACH OTHER OUT!

UH. THAT'S *NOT* HOW IT WORKS.

Some cancers begin with a mutation that directly affects the function of an epigenetic highlighter, eraser or decoder. In these cases, it's a fairly safe bet that the resulting epigenetic changes are a direct cause of the tumour.

The genes that code for epigenetic regulators can be deleted or amplified, shifting the balance of repressive and activating modifications in the cell. The DNA sequences of the genes that code for these regulators can also mutate, changing the sequence – and therefore the function – of the corresponding RNAs and proteins.

Mutated epigenetic regulators can cause large-scale changes to gene activation and protein production patterns.

OH NO, THE HIGHLIGHTERS HAVE GONE ROGUE! THEY'RE TELLING ME TO DIVIDE *AGAIN!*

DOES THAT SAY "MOVE TO THE LIVER"? BUT I'M AN INTESTINAL CELL!

DNA methylation

Mutations that affect every known type of epigenetic regulator have been found in cancer cells, disrupting everything from regulatory RNAs and histones to the proteins that remodel chromatin or add and remove epigenetic modifications.

Epigenetic regulator mutations are particularly common in blood cancers. For example, a single base change to the EZH2 gene is characteristic of several types of lymphoma. The mutated highlighter protein is hyperactive, adding too many methyl groups to histones and inappropriately repressing genes as a result.

Many other similar examples are being found – and as we're about to see, these discoveries are helping scientists to develop new drugs to treat cancer and other diseases.

IN CANCER AND OTHER DISEASES, FINDING A CULPRIT ALSO MEANS FINDING A TARGET – SOMETHING THAT CAN BE TREATED AND FIXED.

Epigenetics in Medicine

The benefits of decades of epigenetics research are now starting to trickle into medical practice. An increasing diversity of drugs and tests based on epigenetic modifications and regulators are entering clinical trials, and some have already been approved for regular use.

Much of the research in this area has focused on anti-cancer drugs that can reverse the abnormal epigenetic modification patterns found in malignant cells. The very abnormalities that make cancer cells so dangerous can also make them vulnerable to drugs that target malignant cells, but not their normal counterparts; cells affected by other diseases have fewer of these chinks in their armour.

CONVENTIONAL CHEMOTHERAPY

EPIGENETIC THERAPY

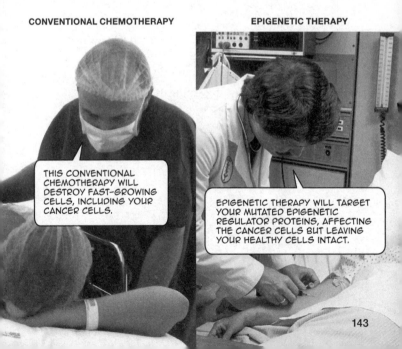

THIS CONVENTIONAL CHEMOTHERAPY WILL DESTROY FAST-GROWING CELLS, INCLUDING YOUR CANCER CELLS.

EPIGENETIC THERAPY WILL TARGET YOUR MUTATED EPIGENETIC REGULATOR PROTEINS, AFFECTING THE CANCER CELLS BUT LEAVING YOUR HEALTHY CELLS INTACT.

There are two possible approaches to reversing the epigenetic changes that take place in cancerous and other abnormal cells. The first is to target the abnormal epigenetic regulators that are ultimately responsible for the changes; the second is to erase and overwrite the modification patterns themselves.

The latter approach is easier and has therefore been better developed, but it has some drawbacks. For example, drugs that remove methyl groups from accidentally silenced parts of the DNA will also indiscriminately remove methyl groups from regions that should stay silent. This lack of specificity can cause drug side effects such as nausea and fatigue. Nevertheless, this non-specific approach does show some promise.

Drugs called **histone deacetylase (HDAC) inhibitors** prevent the removal of acetyl groups from histone tails. They can thus restore the transcription of silenced genes, including tumour suppressors, and have already been approved for use against some cancers. Their benefits might be limited to cancers with very specific types of epigenetic abnormalities, but they are also being tested as treatments for other diseases, including Kabuki syndrome (see page 128).

DNA methyltransferase inhibitors are also being developed. The stable heritability of DNA methylation patterns (see page 47) ensures that the drug's effects will persist in the offspring of any cancer cells that survive the original treatment.

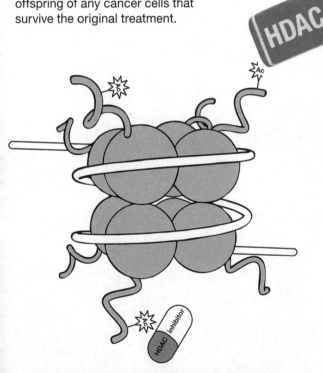

Some progress is being made with HDAC inhibitors and similar molecules, although the benefits haven't been as great as was first hoped. Several drug companies are trying to improve these non-specific therapies, but other researchers are focusing instead on a more specific approach that targets the mutated epigenetic regulators themselves.

This approach mirrors a general trend in cancer therapy, moving away from non-specific drugs that target a general feature of cancer cells – rapid division, or CpG island methylation – towards chemicals that exploit vulnerabilities caused by specific mutated proteins. Targeted therapies cause less collateral damage to healthy cells, and therefore fewer side effects.

NON-SPECIFIC THERAPIES
∨

To illustrate this targeted approach, let's return to the EZH2 mutation found in some types of lymphoma (see page 142). Teams led by **Caretha Creasy** and **Kevin W. Kuntz** have developed chemicals that block the function of the hyperactive mutated EZH2 protein, without blocking its normal counterpart. The drugs haven't been tested in actual cancer patients yet. However, they correct the abnormal pattern of epigenetic modifications and reactivate genes which have been inappropriately silenced in human lymphoma cells grown in the laboratory. Most importantly, the drugs also slow the growth of the lymphoma cells.

Drugs that target other mutated epigenetic regulator proteins which cause cancers are also in development, although it can take decades to develop and approve a new drug. Watch this space!

Therapies based on miRNAs, which can prevent messenger RNAs from being translated into proteins (see pages 66–7), are also being developed. It's possible to create artificial miRNAs that target any mRNA (and therefore any protein) of interest. Potential targets include mutated proteins in cancer cells, the antibodies that cause multiple sclerosis and other autoimmune disorders, the clumps and tangles of proteins that accumulate in Alzheimer's disease, and the viral and bacterial proteins that cause the symptoms of infectious diseases.

No miRNA-based therapies have been approved for regular use yet, and there are concerns about side effects, but this unconventional approach has a lot of promise.

AN EXPERIMENTAL TREATMENT BASED ON MIRNAS WAS TESTED ON EBOLA PATIENTS DURING THE 2014–15 OUTBREAK IN WEST AFRICA, ALTHOUGH PRELIMINARY RESULTS WERE DISAPPOINTING.

Epigenetic modifications can also be used to diagnose certain diseases and to provide physicians with guidance about the best treatment options. Many current tests and treatments are based on proteins: for example, only breast cancers that contain large amounts of HER2 protein can be treated with the drug Herceptin, which specifically kills cells that have HER2 proteins on their outer surfaces.

Epigenetic changes that are specific to certain types of disease can potentially be used in the same way. For example, several companies are developing stool sample tests to detect epigenetic changes that are characteristic of colorectal cancer, as an alternative to existing forms of cancer screening. Similarly, epigenetics-based tests could be used in the future to decide which cancer patients would benefit from HDAC or DNA methyltransferase inhibitor treatments. Epigenetics-based tests aren't as specific as the more conventional alternatives yet, but they're steadily improving.

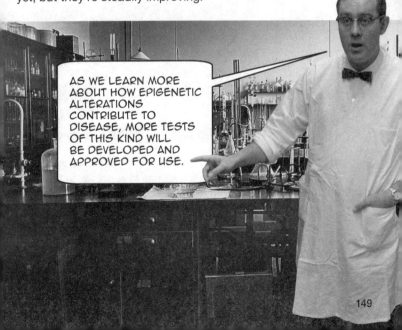

AS WE LEARN MORE ABOUT HOW EPIGENETIC ALTERATIONS CONTRIBUTE TO DISEASE, MORE TESTS OF THIS KIND WILL BE DEVELOPED AND APPROVED FOR USE.

Stem Cell Therapies

A new field of medicine aims to use stem cells – the body's undifferentiated cells from which mature cells arise – to create replacements for damaged mature cells and organs.

STEM CELL THERAPY HAS THE POTENTIAL TO TREAT AND EVEN TO CURE MANY DIFFERENT HUMAN DISEASES.

The differentiation of versatile stem cells into specialized mature cells is usually a one-way process. As we've seen, John Gurdon demonstrated in the 1960s that it's possible to use cloning methods to artificially reverse cell differentiation and create new stem cells (pages 19–20). However, cloning a cell into a whole new embryo in order to extract the clone's stem cells would be an incredibly inefficient way to produce new stem cells for therapeutic purposes – not to mention the ethical problems inherent to human cloning.

As an alternative to reversing the specialization of mature cells, human stem cells can be extracted from very early stage embryos, before they implant into the uterus. Surplus embryos from in vitro fertilization (IVF) procedures are typically used, with the parents' consent. Embryonic stem cells can be grown and differentiated in the laboratory, then transplanted into sick or injured recipients.

Despite some promising animal studies, human trials have been limited by the cells' scarcity and by ethical concerns. There's also a risk that the transplanted cells won't differentiate normally and will become cancerous instead. In addition, because the donor cells would contain different genes and proteins to the recipients' own cells, recipients would need to take immune-suppressing drugs for the rest of their lives to prevent the transplanted cells from being rejected as foreign invaders.

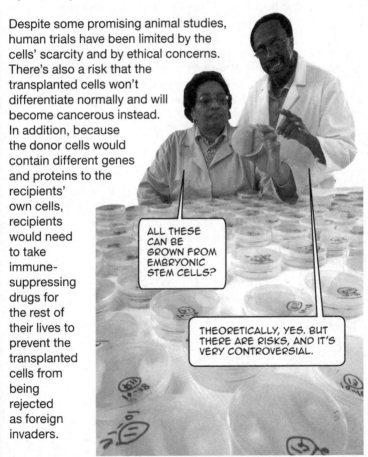

ALL THESE CAN BE GROWN FROM EMBRYONIC STEM CELLS?

THEORETICALLY, YES. BUT THERE ARE RISKS, AND IT'S VERY CONTROVERSIAL.

In 2006, Japanese stem cell biologist **Shinya Yamanaka** (b. 1962) published a new method for creating stem cells directly from mature adult cells. The resulting **induced pluripotent stem cells (iPS cells)*** can differentiate into various mature cell types, just like embryonic stem cells. This means that we could use the patient's own skin or blood cells to repair or replace their damaged tissues and organs, which would remove both the ethical and the immune suppression obstacles that have hampered human embryonic stem cell research.

The initial paper seemed too good to be true, but other teams quickly replicated the results. Yamanaka shared the 2012 Nobel Prize in Physiology or Medicine with John Gurdon, "for the discovery that mature cells can be reprogrammed to become pluripotent".

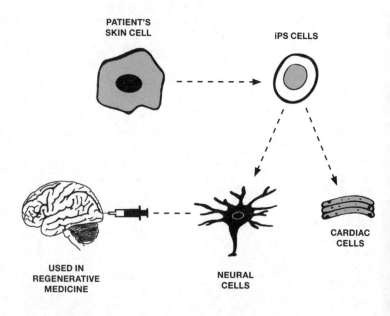

PATIENT'S SKIN CELL

iPS CELLS

USED IN REGENERATIVE MEDICINE

NEURAL CELLS

CARDIAC CELLS

The creation of iPS cells from mature cells is inefficient, in part because the reversal of cell differentiation requires substantial changes to epigenetic modification patterns. For example, natural embryonic stem cells have not yet undergone X chromosome inactivation (see pages 84–6). iPS cells that are created from mature XX cells need to resemble natural embryonic stem cells as closely as possible, and so the silenced chromosome needs to be reactivated – but this process is often left incomplete in XX iPS cells.

Similarly, while iPS cells share many of the unique epigenetic characteristics of the body's natural stem cells, they sometimes also retain methylation patterns that are characteristic of the type of mature cell from which they were derived. This limits their ability to differentiate into other types of cell.

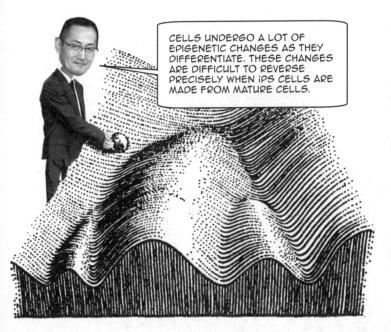

CELLS UNDERGO A LOT OF EPIGENETIC CHANGES AS THEY DIFFERENTIATE. THESE CHANGES ARE DIFFICULT TO REVERSE PRECISELY WHEN iPS CELLS ARE MADE FROM MATURE CELLS.

Yamanaka created the first iPS cells by inserting four transcription factors into mature cells. Three have known epigenetic functions: Oct4 blocks histone methylation, while c-Myc and Klf4 bind to histone acetylating proteins. There's evidence that the fourth transcription factor, Sox2, is also associated with histone modification changes.

Because excess amounts of c-Myc and Klf4 can cause cancer, alternative iPS cell production methods are needed. Two American teams, led by **Sheng Ding** and **Douglas Melton** (b. 1953), showed in the late 2000s that inhibitors of DNA methylation, histone methylation or histone deacetylation can substitute for some of the four transcription factors. However, the epigenetic changes caused by these drugs might also prime the cell to become cancerous.

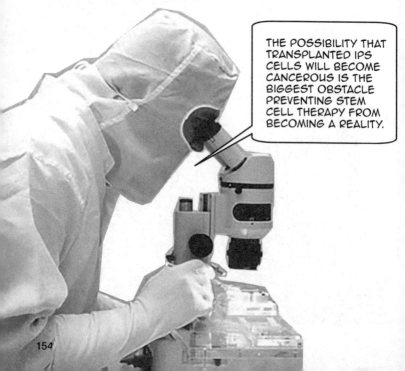

THE POSSIBILITY THAT TRANSPLANTED IPS CELLS WILL BECOME CANCEROUS IS THE BIGGEST OBSTACLE PREVENTING STEM CELL THERAPY FROM BECOMING A REALITY.

There have been some promising animal tests using iPS cell therapy to treat spinal cord injuries and other conditions. Japanese stem cell researchers **Takanori Takebe** and **Hideki Taniguchi** have also managed to grow pieces of liver from human iPS cells.

The first human clinical trial using iPS cells – to treat an eye disease called macular degeneration – began in Japan in late 2014, led by ophthalmologist **Masayo Takahashi**; other trials will no doubt follow if the first proves safe.

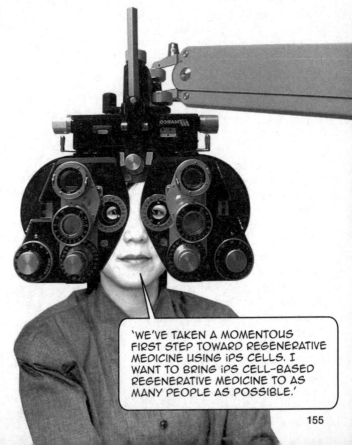

'WE'VE TAKEN A MOMENTOUS FIRST STEP TOWARD REGENERATIVE MEDICINE USING iPS CELLS. I WANT TO BRING iPS CELL-BASED REGENERATIVE MEDICINE TO AS MANY PEOPLE AS POSSIBLE.'

Epigenetics and Pseudoscience

REVERSE THE HARMFUL EPIGENETICS YOUR GRANDMOTHER GAVE YOU!

While exciting progress is being made with the epigenetics-based therapies described earlier, it can take decades to translate a scientific finding into tangible medical advances. It's not surprising that people look outside the scientific mainstream during that lag time.

Epigenetics seems to attract more than its fair share of hype, over-interpretation and pseudoscience. The idea that there's more to us than just our DNA sequence is a very compelling one, after all. The suggestion that we might be able to seek out environmental factors that will override our genetics and cancel out harmful exposures and experiences – including those we might have inherited from our parents and grandparents – is particularly attractive. It's very tempting to believe that these protective environmental factors can be purchased as a nutritional supplement or in another convenient format.

Epigenetics has been touted as the science behind everything from homeopathy and acupuncture (where the treatments are said to cause epigenetic modification changes that improve health) to past life regression hypnosis (where findings from epigenetic inheritance studies, such as those of the Dutch Hunger Winter and Överkalix harvest records, are extrapolated wildly to encompass the transmission of very specific memories of past lives).

However, there's no credible evidence to support these claims. Even with substances that we do know to affect epigenetic modification patterns, we're nowhere near being able to say: "this herbal supplement will silence this specific harmful gene you've inherited". And specific genes can't be epigenetically regulated by the power of thought alone, as has been claimed. Sorry.

CONCENTRATE HARD AND PROJECT YOUR THOUGHTS TOWARDS THE BLOOD CELL WHICH IS MAKING THE MUTATED PROTEINS. THEN YOU CAN ACTUALLY TELL IT TO STOP.

IT'S A NICE IDEA, BUT IT'S NOT REMOTELY TRUE.

The Future of Epigenetics

The fast pace of discovery and broad applications of epigenetics make it an attractive field for early-career scientists, with lots of room to carve out a unique research niche.

Epigenetics is a young field, and we still have a lot to learn about how the components of the epigenetic regulation network interact with each other to control gene activity. None of the research outlined so far in this book should be considered complete; work is still ongoing in all subfields, and historical precedent predicts that we will have to revise some of our current understanding of epigenetics as we learn more about it.

Epigenomics

As in many other fields of biology, recent advances in DNA sequencing technology have accelerated the pace of epigenetics research. Bisulphite (DNA methylation) and ChIP (histone modification) sequencing, combined with RNA sequencing, now allow the study of the **epigenome*** – epigenetic modifications on the whole-genome scale.

The International Human Epigenome Consortium (IHEC) and other collaborative projects are engaged in generating human epigenome sequences for thousands of normal and diseased cell types. The data are made available online and are being used by research groups around the world to further our understanding of diverse aspects of epigenetics, from gene regulation to disease susceptibility to evolution.

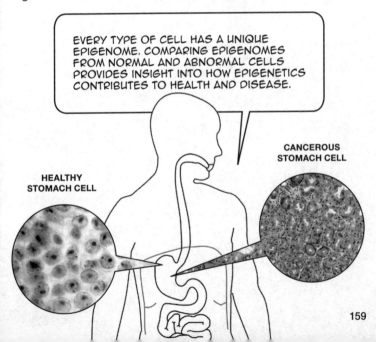

EVERY TYPE OF CELL HAS A UNIQUE EPIGENOME. COMPARING EPIGENOMES FROM NORMAL AND ABNORMAL CELLS PROVIDES INSIGHT INTO HOW EPIGENETICS CONTRIBUTES TO HEALTH AND DISEASE.

CANCEROUS STOMACH CELL

HEALTHY STOMACH CELL

One of the biggest challenges in epigenomics is the huge number of cells that are needed for bisulphite and ChIP sequencing. These requirements limit the types of tissue that can be sequenced, especially for normal cells; it's relatively easy to obtain large surgically-removed tumours, for instance, but much harder to obtain enough matched normal tissue from living donors to use as a control.

Public and private sector teams are working on developing new sequencing methods that can process smaller samples. Some labs are even working on methods to study DNA methylation in single cells, which is more informative than current methods that assess average values across many cells.

New Epigenetic Modifications

The list of known histone modifications is constantly being expanded by the discovery of new types of modification and of known modifications in new locations. New types of DNA modification are being discovered, too. For example, in addition to methylated C bases, there are molecular variants called hydroxymethylcytosine (hmC), formylcytosine (fC) and carboxylcytosine (caC). hmC was already known to be involved in active DNA demethylation; an American team led by **Yi Zhang** has shown that fC and caC play similar roles. In 2015, a British group led by **Shankar Balasubramanian** found that fC can also physically force the DNA double helix into a more open configuration. We still know very little about these variants, but learning more about them will help us better understand how and why methylated DNA is sometimes reactivated.

Three 2015 papers by **Chuan He**, **Yang Shi**, **Hailin Wang** and **Dahua Chen** demonstrated that the A bases of DNA can be methylated in algae, flies and worms. **Methylated A bases** had already been found in single-cell bacteria, where their functions include DNA repair and protection against viruses. However, the 2015 studies were the first evidence that this epigenetic modification also occurs in multicellular species – and that it might therefore exist in humans. While we know very little indeed about the role of methylated A bases in epigenetics, there's preliminary evidence that – like methylated C bases – they're involved in cell differentiation and can interact with histone modifications.

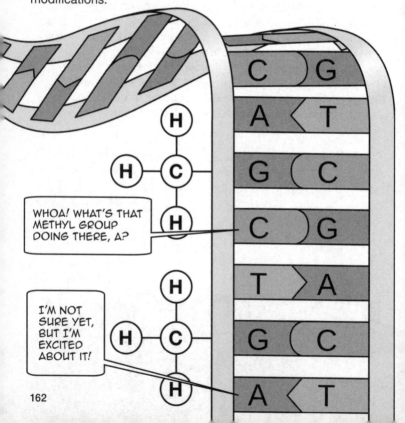

The Epitranscriptome

The bases that make up RNA can also be modified. In fact, there are more than 100 types of RNA base modification. Methylation of the A bases seems to be the closest equivalent to DNA C-base methylation: methylation is reversible, stem cells have characteristic methyl-A patterns, and the pattern can change in response to environmental cues.

As yet there's no equivalent of bisulphite sequencing to directly identify individual methylated A bases. However, techniques similar to those used to map the locations of histone modifications have yielded some interesting preliminary findings about which RNA strands tend to be methylated.

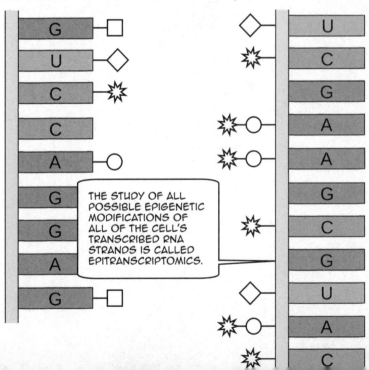

THE STUDY OF ALL POSSIBLE EPIGENETIC MODIFICATIONS OF ALL OF THE CELL'S TRANSCRIBED RNA STRANDS IS CALLED EPITRANSCRIPTOMICS.

Methylation modifies mRNA stability, which in turn determines the abundance of the corresponding protein. The possibility of other functions was raised by the discovery that some miRNAs and proteins can bind only to methylated (or only to unmethylated) RNA strands. There's indirect evidence that RNA A-base methylation plays a role in determining which pieces of the mRNA precursor strand are spliced together to code for the final protein; the other consequences of methylation-specific RNA binding aren't yet known.

I GUESS THIS METHYL GROUP MAKES ME MORE STABLE?

Epigenetic Editing

The epigenetics-based drugs used to fight cancer and other diseases are still relatively blunt tools. Some can alter the total amount of DNA methylation or histone acetylation in the cell, while others can inhibit proteins that epigenetically regulate multiple target genes (see pages 144–8). However, the ability to selectively regulate a specific gene is beyond the reach of conventional pharmaceuticals.

An emerging alternative approach uses genetically engineered hybrid proteins to reactivate or silence an individual gene. The hybrids are made by fusing part of a transcription factor, which can bind to specific sequences in the genome, to a protein that can add or remove DNA methylation or histone modifications.

New epigenetic editing techniques can change the way the script of our DNA sequence is read and interpreted, with the potential to reverse the harmful epigenetic modification patterns that are common in cancer and other diseases. A new DNA sequence editing technique called CRISPR can be adapted so that it edits epigenetic modifications instead of editing the DNA sequence directly. Part of a protein called Cas9 is used instead of a transcription factor's DNA binding domain. Cas9 binds to a "guide" RNA strand that seeks out complementary DNA sequences; scientists can insert any sequence of bases into this RNA strand. Hybrid epigenetic regulator proteins that contain Cas9 domains can thus be directed to any DNA sequence of interest, allowing a specific gene to be activated or repressed. Epigenetic editing has only been demonstrated in cultured cells so far, and will have to undergo rigorous safety testing before it can be used in humans. If proven safe, this technology has the potential to treat cancer and other diseases that involve changes in gene activation patterns.

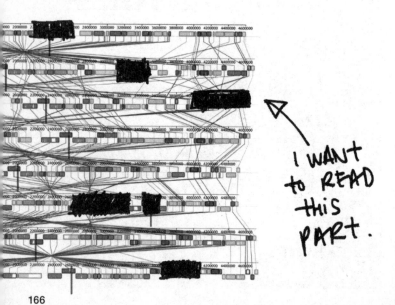

I WANt to REAd this PARt.

Epigen-Ethics

Studying epigenetics creates unique ethical challenges around potential breaches of patient privacy, discrimination on medical grounds, law enforcement, parenting and the risk of passing on the effects of harmful experiences to new generations.

Scientists generally make their sequencing data available to other researchers; in fact, they're often required to do so. There's a risk therefore that individuals who donate cells for research could be identified from their epigenetic data, with implications for their privacy and (in some countries) health insurance coverage.

The first couple of generations of DNA sequencing technology could only read short fragments of genes. It was generally believed that as long as the donor's name and demographic details were stripped from the sequencing data, the sequences themselves were essentially anonymous.

I WAGER YOU CAN'T TELL WHICH DNA BELONGS TO WHICH OF US!

WELL, I SEE A Y CHROMOSOME SEQUENCE, SO TO START WITH, I'M GOING TO SAY IT'S FROM A MAN. AND THAT'S JUST THE BEGINNING ...

As technology improves, the amount of sequencing data generated is increasing exponentially – and more and more information can be inferred from the sequences. We've learned that certain DNA sequences are only ever found in people whose ancestors came from specific geographical regions. Individuals can also now undergo private DNA sequencing, to learn more about their health risks or ancestry. Some customers choose to associate their DNA sequence data with their real names and family trees.

The convergence of these and other data types allowed American computer scientist **Yaniv Erlich** to demonstrate in 2013 that it's possible to infer the surnames of some donors from parts of their DNA sequence alone.

It's harder to identify individuals from their epigenomes than from their genomes: bisulphite sequencing changes the identity of some bases, while histone modification and RNA sequencing methods only capture short fragments of the genome. However, as computer algorithms improve and more data become available, it's becoming possible to identify participants in epigenetic research.

Epigenetics researchers need access to detailed metadata. Epigenetic modification patterns change with age and in certain diseases; researchers therefore often need to collect and publish the approximate age, health status and other details of the donor in order to make sense of their epigenome data. Making these very personal metadata available along with the sequence data increases the privacy risks for research participants.

THESE EPIGENETIC MODIFICATION PATTERNS ARE CHARACTERISTIC OF AN OLD EUROPEAN MALE.

WE'RE BOTH VERY, VERY OLD.

The ability to identify the epigenetic traces of some environmental exposures might also have privacy implications (see page 167). It's not yet possible to study an epigenome and say "this person eats too much saturated fat" or "this person used to use cocaine", but it might be in the future. Without regulations in place, epigenetic data could therefore potentially be used for criminal profiling, or to deny people employment or health insurance.

We're still grappling with many of the ethical conundrums involved in genetics research, let alone epigenetics. Scientists and ethicists continue to study and debate these issues, which are evolving along with our understanding of the science.

LE VOYAGE DU *H.M.S. BEAGLE*
1831–1836

AHA! EPIGENETIC MODIFICATION PATTERNS CHARACTERISTIC OF A LONG SEA VOYAGE. THIS IS CHARLES DARWIN'S EPIGENOME.

IMPRESSIVE! I'M GLAD I HAVEN'T COMMITTED ANY CRIMES IN THE PRESENCE OF ALL THIS NEW TECHNOLOGY.

Looking Ahead

Scientists like to joke that epigenetics can and will explain everything. Unfortunately, some people take the joke a little too seriously: the field of epigenetics as a whole is particularly susceptible to over-interpretation, unrealistic hype and even deliberate misrepresentation. Experts and the public alike therefore need to exercise a degree of caution when it comes to claims about epigenetics.

IT'S HARD NOT TO GET CARRIED AWAY BY THE POSSIBILITIES!

However, the joke that epigenetics can explain everything didn't appear from thin air; epigenetics has already explained a great deal, filling in many gaps in our knowledge of genetics and other scientific subjects – from embryonic development to genetic regulation, inheritance, evolution and disease. Scientists working in diverse fields, in laboratories all over the world, have made incredible advances in the few decades since Conrad Waddington first coined the word "epigenetics".

By identifying and studying the chemical changes that affect how the raw DNA sequence is used to produce RNAs and proteins, we have developed a better understanding of all aspects of biology – and we're starting to learn how to use this knowledge to improve human health.

Glossary

Acetyl group: A small molecule with the chemical formula $COCH_3$ and a negative charge. Acetyl groups can be attached to **histones** and other **proteins**, changing the protein's shape and/or charge and thus affecting its functions.

Amino acid: A molecule that serves as the building block of **proteins**. There are 20–23 types of amino acid, depending on species. One part of the molecular structure is different in each type of amino acid; this part determines the charge and chemical properties of the molecule, such as whether it's attracted to or repelled by water. The sequence of amino acids in a protein determines the shape, charge, chemical properties and functions of the protein.

Cell: A self-contained living entity, which makes up humans and other complex organisms. There are many different specialized types, which are packed together to form tissues and organs. Each cell is surrounded by a membrane that controls which molecules can move in and out; it contains distinct structures, such as the **nucleus** and mitochondria (which produce the cell's energy).

Cell differentiation: The process by which a versatile **stem cell** produces the body's specialized mature cells. It occurs through several rounds of mitosis: cells become gradually more specialized with each division until they are fully differentiated and can only produce copies of themselves.

Cell division: A process during which a single cell divides into two. It is needed to replace dying cells that become damaged, infected with a virus, or reach the end of their lifespan. Excessive cell division can cause problems, including cancer; cell division is therefore very carefully controlled.

Chromatin: A combination of DNA and the histones, scaffolding proteins, other proteins and RNAs that bind to it.

Chromosome: The long strands of chromatin that comprise a **genome**. Different species possess different numbers of chromosomes.

Codon: A sequence of three consecutive bases in an **mRNA** strand. Each codon specifies a single **amino acid**, although some amino acids are encoded by multiple codons.

CpG island: A cluster of CpGs (a C base adjacent to a G base, linked by a phosphate group, p) close to the transcription start site of a **gene**. The 70% of genes that have a CpG island are generally more active than the 30% that lack such structures.

Deoxyribonucleic acid (DNA): The molecule that carries genetic information from one generation to the next. DNA comprises long strings of molecules called nucleotides, each of which has a base, a sugar molecule called deoxyribose, and a phosphate group. The deoxyribose and phosphate molecules are identical in every nucleotide, but there are four different bases – adenine (A), cytosine (C), guanine (G) and thymine (T). Two strands of DNA with "complementary" sequences coil around each other in a double helix, with the bases pairing on the inside and the sugar and phosphate groups on the outside. A always pairs with T, and C with G.

Epigenetic reprogramming: The removal and subsequent reestablishment of DNA methylation in embryonic cells. It first occurs early in embryonic development (week one) and affects every cell; the second reprogramming is restricted to the **primordial germ cells** and occurs in weeks ten to eleven.

Epigenome: The total combined epigenetic modifications of the entire **genome**. Unlike the genome, which is essentially identical in every cell of an individual organism, the epigenome differs in different cell types, under different conditions.

Gene: A section of **DNA** that codes for a specific **RNA** and/or **protein**. Genes usually consist of promoters (regulatory regions), exons (which code for amino acids), and introns (which are spliced out of the mRNA before protein translation begins). Genes sometimes overlap, so the same piece of DNA can contain more than one gene.

Genome: The total DNA of a given cell or species. The human genome was sequenced by the Human Genome Project, completed in 2003.

Heritability: The extent to which variation in a trait is determined by genes, usually expressed as a percentage. Most conditions are controlled by more than one gene and are affected by environmental factors, thus have a heritability score below 100%.

Histone: A class of **protein** that binds closely to DNA, helping to pack it into the **nucleo-some** structures that form the basic unit of **chromatin**.

Histone variants: Non-standard versions of **histone** proteins, which have specialized functions and bind to DNA only under certain conditions, such as when the DNA is damaged and needs to be repaired.

Imprinting: The epigenetic phenomenon by which some **genes** are transcribed from only the maternal, or only the paternal, **chromosome**, in some or all cells.

Induced pluripotent stem cell (iPS cell): A **stem cell** that has been artificially created from a fully differentiated mature cell in a laboratory.

Long non-coding RNA (lncRNA): An **RNA** strand at least 200 bases long that doesn't code for a protein. Among various functions, lncRNAs act as guides to connect proteins to each other and to specific locations in the **genome**; determine which epigenetic modifications should be added to which pieces of DNA; and mop up **miRNAs** so that they can't bind to mRNAs.

Meiosis: The specialized type of **cell division** that creates egg and sperm cells. Meiosis consists of DNA replication followed by two rounds of cell division, creating four cells that each contain half the amount of genetic material as the original cell. Genetic recombination between paired **chromosomes** occurs before the first cell division, shuffling the genes to give each egg or sperm cell a unique **genome**.

Messenger RNA (mRNA): An RNA strand that codes for a **protein**. A molecular cap and a string of A bases help the translation machinery convert the sequence of mRNA codons into the corresponding **amino acid** sequence.

Methyl group: A small molecule with the chemical formula CH_3 and a neutral charge, which can be attached to DNA, histones and other proteins (a process known as methylation), changing the molecule's shape and thus affecting its functions. DNA methylation silences gene transcription.

MicroRNA (miRNA): A 19–24 base RNA strand that can bind to complementary sequences within mRNAs and that prevents protein **translation** by either physically blocking the translation machinery from accessing the mRNA or by triggering the destruction of the mRNA strand.

Mitosis: Standard type of **cell division**, consisting of DNA replication followed by a single cell division, creating two cells that each contain the same amount of genetic material as the original cell. No genetic recombination occurs.

Molecule: A discrete chemical structure formed by a combination of atoms of the same or different types.

Nucleosome: The basic unit of **chromatin**, consisting of DNA coiled around 8 **histone** proteins. Adjacent nucleosomes are linked by a DNA sequence bound by a single protein.

Nucleus: A compartment inside most **cells**, surrounded by a membrane and containing the cell's chromosomes plus certain proteins and RNAs. Different regions within the nucleus specialize in different functions, such as gene transcription.

Phenotype: The observable characteristics of an organism, such as height, eye colour and hair colour. The visible result of an individual's combination of genes (their genotype) and environmental influences.

Primordial germ cells: The cells within the developing embryo that, after cell differentiation, produce the egg or sperm cells.

Protein: A large molecule comprising a string of **amino acids** with diverse functions: some are structural, others perform and control chemical reactions, others fight infection. The protein's unique shape and chemical properties determine which molecules it can bind to and what functions it can perform.

Receptor protein: A protein that binds specifically to one type (or family) of molecule. It usually changes shape when it binds to this molecule, triggering downstream changes in protein-protein binding and other phenomena. There are receptor proteins for external molecules, such as those absorbed from food and the environment, and for internally-produced molecules, such as hormones.

Recombination: A process by which, prior to the first cell division during **meiosis**, the parents' maternally- and paternally-derived chromosomes swap pieces of DNA, resulting in a hybrid **chromosome** which is passed to the resulting egg or sperm cell.

Repetitive DNA: Short DNA sequences that appear multiple times in the genome. Some repeats cluster together, while others are interspersed throughout the genome. Most repetitive DNA is silenced by epigenetic mechanisms, but some escape these controls, because their functions are useful to the cell, through chance or because they are reactivated in cancerous and other abnormal cells.

Ribonucleic acid (RNA): RNA is similar to **DNA**, except that ribose, instead of deoxyribose, is the sugar component of each nucleotide; the uracil (U) base is used instead of T; and it can exist as a single strand. RNAs serve many functions in the cell, from messenger RNAs to regulatory RNAs.

Signalling cascade: A molecular process by which the binding of a molecule to the corresponding **receptor protein** triggers a series of reactions, resulting in a change in gene activation patterns. The signal is passed from protein to protein via molecular changes to the shape and/or function of each protein in the cascade.

Stem cell: An undifferentiated **cell** with the ability to produce different types of differentiated cells. The most versatile stem cells ("totipotent") exist in the earliest-stage embryo and can produce all the mature cells of the embryo's body plus the placenta; "pluripotent" stem cells can produce all cells except the placenta; "multipotent" stem cells can produce a limited set of cell types within one specific type of tissue – e.g. multiple types of blood cell. When a stem cell divides it typically creates one exact copy of itself (a process called self-renewal) and one cell that will differentiate into either a less versatile stem cell or a mature cell. They are active throughout embryonic development and into our adult lives.

Transcription: The production of a complementary RNA strand from a DNA template.

Transcription factor: A protein that interacts with DNA and other proteins to either activate or repress transcription. Some TFs are are necessary for the transcription of every gene; others are produced only in certain cell types, at particular stages of embryonic development, or in response to specific **signalling cascades**.

Transfer RNA (tRNA): Involved in the **translation** of mRNAs into proteins, each tRNA strand is about 75–90 bases long and folds into a cloverleaf secondary structure.

Translation: The process by which the gene's code, carried in the **mRNA** strand, is converted into a specific sequence of **amino acids**, producing a **protein**.

Zygote: The single **cell** produced when an egg cell is fertilized by fusing with a sperm cell. The zygote divides by mitosis to produce a two-cell embryo.

Recommended Further Reading

Introducing Genetics: A Graphic Guide by Steve Jones and Borin van Loon (Icon Books, 2011)

Introducing Evolution: A Graphic Guide by Dylan Evans and Howard Selina (Icon Books, 2001, 2010)

The Epigenetics Revolution: How Modern Biology Is Rewriting Our Understanding of Genetics, Disease and Inheritance by Nessa Carey (Icon Books, 2012)

Herding Hemingway's Cats: Understanding How Our Genes Work by Kat Arney (Bloomsbury Sigma, 2016)

The Emperor of All Maladies: A Biography of Cancer by Siddhartha Mukherjee (Fourth Estate, 2011)

https://www.coursera.org/course/epigenetics: Online course. Requires an understanding of the basics of genetics.

http://ihec-epigenomes.org/why-epigenomics/: Collection of articles and videos for a non-specialist audience.

Author's Acknowledgements

Many thanks to Kiera Jamison, Oliver Pugh and all at Icon Books for the opportunity, and for making this book happen! I'd also like to thank Catherine Anderson for her excellent advice (and wine).

I'm also grateful to David Gillespie and Dixie Mager for teaching me so much about research and writing, and to Martin Hirst, Marco Marra, Steve Jones, İnanç Birol, Sam Aparicio and David Huntsman, plus Dominik Stoll, Joanne Johnson, Robyn Roscoe and the rest of the GSC Projects team, for giving me the opportunity to work on super cool genomics and epigenomics projects.

Thanks also to my fellow IHEC Communications working group members for making those 5:30am teleconferences fun (after the tea kicks in); Eva Amsen, Erika Cule, Stephen Curry, Henry Gee, Richard Grant, Bob O'Hara, Frank Norman, Jenny Rohn, Steffi Suhr, Richard Wintle and the rest of the Occam's Corner/Occam's Typewriter blog collective (and friends) for all their support and helpful writing advice over the years; ditto Jane O'Hara, Anne Steinø and Susan Vickers; ditto the Just Write Vancouver Saturday Morning Meetup gang.

I'd also like to thank Ann, Tom and Claire Dunn, and my wonderful husband Mark Ennis and all his family, for their years of love and support.

Cath Ennis has a research background in genetics, genomics and cancer, and works as a grant writer and project manager in Vancouver, Canada. She can be found online at www.enniscath.com.

Oliver Pugh is a designer and the illustrator of *Introducing Infinity* and *Introducing Particle Physics*.